Download the TotalAR app to access enhanced features including video solutions and the updated NCEES *FE Reference Handbook* page numbers for the solution references.

Scan this QR code with a smartphone or tablet to download the TotalAR app.

Available on the App Store

GET IT ON Google Play

Once you have downloaded the app, scan this TotalAR code to activate the AR features in the book:

When the code is recognized, you will be prompted to enter your book's unique alphanumeric access key:

Unique Access Key:
MXA4EP53

School of PE™
A Division of EDUMIND

www.schoolofpe.com

FE Mechanical Practice Exam & Solutions

EduMind, Inc. is not affiliated with the National Council of Examiners for Engineering and Surveying (NCEES) and all questions are not actual exam questions or questions provided by NCEES, but are similar in nature to the type of questions expected on the exam. All questions and provided solutions are not guaranteed to be error free and the sole intended use of the questions and solutions is for use as a study aid and not for any practical application.

For more details on the NCEES FE Mechanical exam, visit https://ncees.org/engineering/fe/.

January 2022

ISBN 978-1-970105-44-5

Printed in the USA

School of PE
An imprint of EduMind, Inc.
425 Metro Place N, Suite 450, Dublin, OH 43017
www.schoolofpe.com

Introduction

School of PE takes test preparation seriously, and we want users to pass their FE Mechanical exams. This *FE Mechanical Practice Exam & Solutions* volume contains enhanced features to help users gain an edge in preparing for the FE Mechanical exam.

We know that taking practice exams is a big part of the test-prep process. Sometimes, it helps to have someone explain the solutions step by step. That's why this volume includes brief videos with each solution, in which a subject-matter expert walks the reader through the problem solution steps.

Another big part of preparing for the FE Mechanical exam is learning to navigate the NCEES *FE Reference Handbook (FERH)*, the only resource test-takers can access during the exam. In most of this volume's solutions, you will see an NCEES *FERH* section reference, directing the user to the section of the *FERH* that covers the relevant topic. (The questions that don't have an *FERH* reference were deemed by our subject-matter experts to be basic knowledge for FE Mechanical test-takers.) You will notice that we did not include *FERH* page numbers in the volume, as these change with each new version of the handbook. By looking for the necessary information in the *FERH*, you will learn how it is organized, saving you time during the exam. To find the *FERH* page numbers of the references that appear in this book, use the augmented reality (AR) feature—*FERH* Content Locator. The page numbers will be updated with the latest version of the handbook so you will always have the most up-to-date information handy.

To access these features, download the TotalAR app (see the first page of this volume). This will allow you to gain access to the book's enhanced content. When you see a TotalAR (TAR) code, like the one for the *FERH* Content Locator (p. iv), simply scan it with the TotalAR app to access the desired content. When you scan your first TAR code in the book, you will be prompted to enter a unique access key, which you will also find on p. i.

With this book, you also have access to an online version of this practice exam via the School of PE Exam Simulator, available on www.schoolofpe.com. If you purchased the book through our website, you should have received an email with a user ID and password. Once you log in, click the icon for this book to gain access to the online version of this practice exam. If you purchased the book from another site—for example, Amazon—please email your name and unique access key (from p. i) to info@schoolofpe.com to gain access to this exclusive online feature.

A Final Note

We have tried our best to make this book error free; it has been technically reviewed, edited, and tested. However, we are only human, and errors can happen. If you spot one, please notify us so we can improve the content. To report an error, scan the Feedback code (below) with the TotalAR app, or visit this feedback page: https://publications.schoolofpe.com/books/fe-mechanical-practice-exam/feedback.

We will verify the correct information and add it to the errata page, which will be made available immediately via the TAR code below and the following link: https://publications.schoolofpe.com/docs/fempes/peb/v2/errata.pdf.

We welcome your feedback and want to hear how well this book prepared you for the FE Mechanical exam. If you have any suggestions or comments, please send them to us at publications@edumind.com.

Best of luck!
The School of PE team

***FERH* Content Locator** **Feedback** **Errata**

PRACTICE EXAM

1. Beginning at point A, mark a point on the vertical axis that would result in a line with a slope of 4/-3.

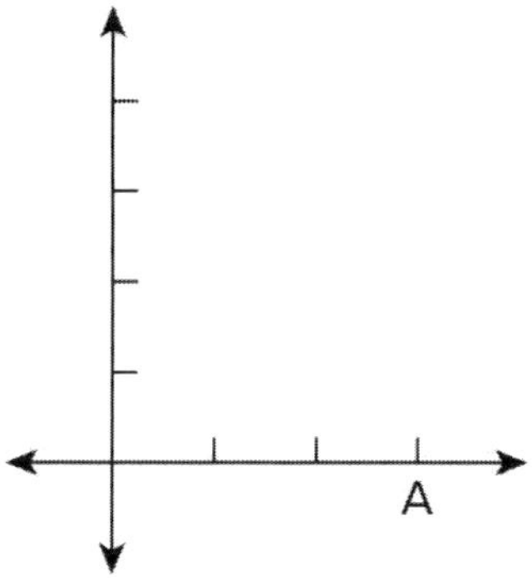

2. Find the correlation between the given set of data:

X	60	61	62	63	64	65	66
Y	3.1	3.6	3.8	4	4.1	4.6	6

A. 0.91
B. 1.5
C. 2.1
D. 2.3

3. With a 10% interest rate, compute the present equivalent.

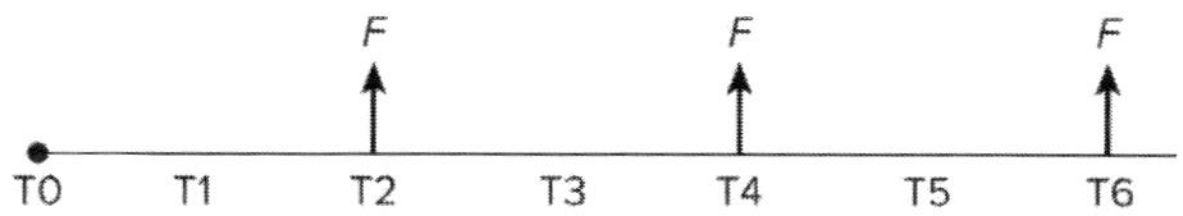

Consider F = \$100 and T1 = 1, T2 = 2, T3 – 3, T4 = 4, T5 – 5, T6 = 6

A. \$47.62
B. \$123.54
C. \$207
D. \$412

4. A concrete block is placed on a wooden plank. The coefficient of static friction between the block and plank is 0.35. The angle between the wooden plank and horizontal surface was originally zero. Gradually, the plank is lifted. Determine the angle at which the block begins to slide.

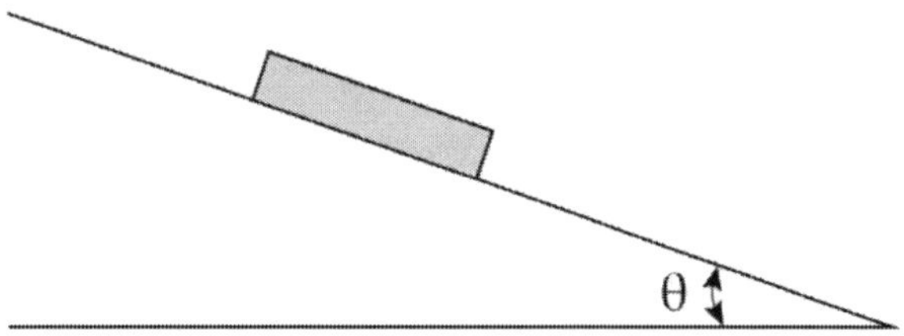

A. 19.3°
B. 19.7°
C. 20.3°
D. 20.8°

5. Determine the normal stress value corresponding to the center of the Mohr's circle representing the 2D stress element shown below.

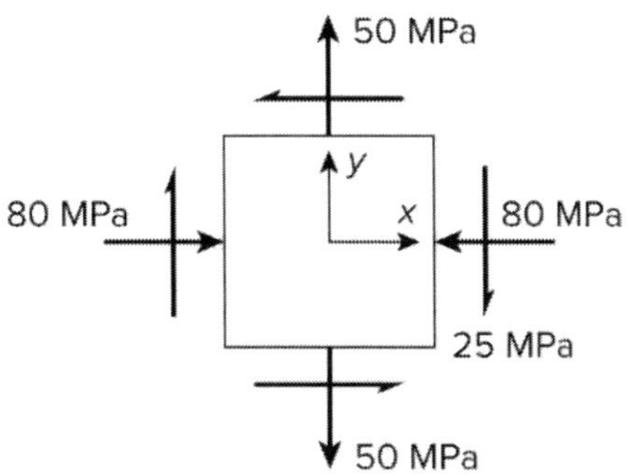

A. -15 MPa
B. 15 MPa
C. -65 MPa
D. 65 MPa

6. Which of the following are subcategories of natural and forced vibrations? Select as many as apply.

A. Damped
B. Undamped
C. Rectilinear
D. Linear
E. Angular

7. What is the voltage across the 10-Ω resistor?

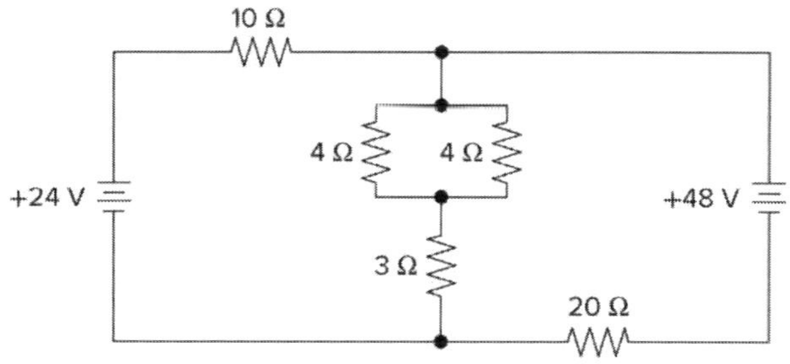

A. 4.8 V
B. 10.3 V
C. 12.2 V
D. 24 V

8. Which of the following are rules in the NCEES Rules of Professional Conduct? Select all that apply.

A. Do not partner or practice with those engaged in dishonest business practices.
B. Safeguard the health, safety, and welfare of the public when performing client work.
C. It is accepted practice to issue statements paid for by interested parties without revealing personal interest in the matter.
D. Do not misrepresent academic or professional qualifications.

9. Suppose that 5% of all light bulbs produced by a particular manufacturer are defective. If we buy a package of eight light bulbs, what is the probability that one is defective?

A. 0.239
B. 0.279
C. 0.297
D. 0.721

10. Considering the following equation of line $y = 7x + 2$, which statement is accurate?

A. -2/7 is the x intercept of the line.
B. -2/7 is the y intercept of the line.
C. 2 is the slope of this line.
D. 2 is the x intercept.

11. Which of the following is the correct relationship between a capital recovery factor $(A/P, i\%, n)$ and a uniform series sinking fund factor $(A/F, i\%, n)$?

A. $(A/P, i\%, n)/i\% = (A/F, i\%, n)$
B. $(A/F, i\%, n)/i\% = (A/P, i\%, n)$
C. $(A/P, i\%, n) + i\% = (A/F, i\%, n)$
D. $(A/F, i\%, n) + i\% = (A/P, i\%, n)$

12. An engineer has a professional engineering license only in civil engineering. Which of the following is true?

A. They can always stamp mechanical drawings because mechanical and civil are under the same umbrella.
B. They can stamp electrical drawings if they have a friend, who is an electrical engineer, proofread the drawings.
C. They can stamp a civil drawing they did not work on or oversee the work on.
D. None of the above

13. An ideal servomotor has an applied motor voltage constant of K_E = 1.3505 V/rad/s. What is the torque created from 40A and 150 Ω at 150 rpm?

A. 26 N·m
B. 47 N·m
C. 54 N·m
D. 75 N·m

14. A cantilevered beam is subject to a bending moment at the free end as shown below. Which of the following statements are correct? Select all that apply.

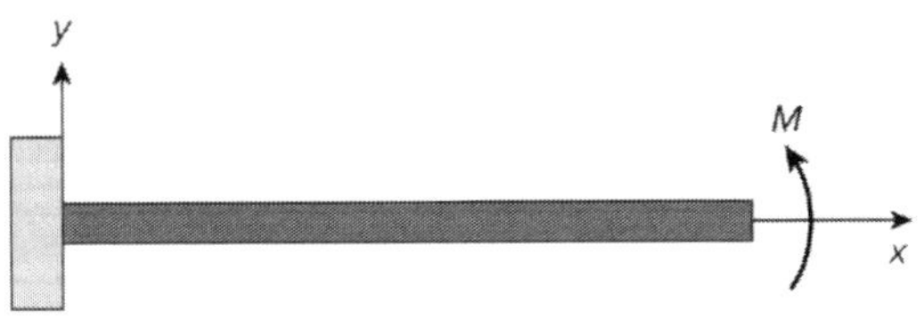

A. The slope at the fixed end is zero.
B. The slope at the free end is the maximum, and it is positive.
C. The slope at the free end is the maximum, and it is negative.
D. The slope at the midpoint is negative.
E. None of the statements are correct.

15. If a system is subjected to a suddenly applied nonperiodic force, the response will be:

A. periodic.
B. steady.
C. transient.
D. None of the above

16. A rope is looped around a pulley as shown below. The pulley is a perfect circle, and the rope extends from the pulley at an angle to the horizontal as shown. The input force (F2) is 120 lb and the coefficient of friction between the rope and pulley is 0.22. Determine the output force (F1) for the system.

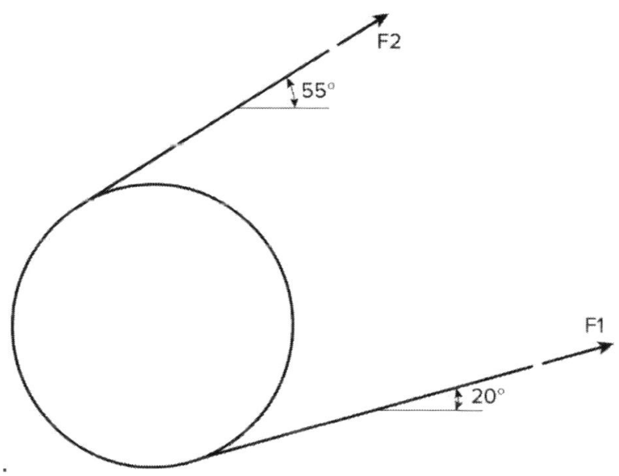

A. 160 lb
B. 210 lb
C. 285 lb
D. 330 lb

17. A commercially available, multiplexing data logger has a switched precision voltage excitation that can supply +2.5 VDC at up to 25 mA of current. A schematic of the compatible bridge completion modules for 350-Ω strain gauges is shown below. Ignoring changes in resistance of the strain gauge and assuming that the data logger inputs have infinite impedance, the maximum number of 350-Ω strain gauge modules that can be driven from the switched voltage excitation without exceeding the rated current is most nearly:

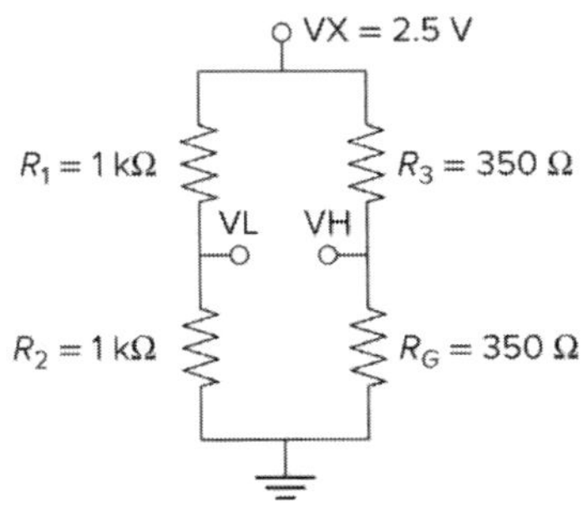

A. 3
B. 4
C. 5
D. 6

18. Annual maintenance costs for a bridge are $3,000. If a new layer of asphalt were placed on the bridge, the annual maintenance would decrease to $1,000 for the first five years, $2,000 for the next five years, and then back to $3,000 after ten years. What is the maximum investment that can justify adding the new layer of asphalt with an interest rate of 6%?

A. $3,129
B. $11,573
C. $15,785
D. $18,933

19. What must be known to calculate the sample coefficient of variation? Select all that are required.

A. Sample standard deviation
B. Arithmetic mean
C. Sample range
D. Sample mode

20. Suppose a random variable X is uniformly distributed on the range [3, 6]. Let $f(x)$ be the probability density function of X. Determine $f(4)$.

A. 1/3
B. 1/4
C. 1/2
D. 1/6

21. A licensed engineer is testing water pipe pressure and realizes one of the lines is not holding the required pressure, but only by a couple of psi. What should they do?

A. Say nothing; a couple of psi will not make a difference.
B. Say nothing; it would cost the company too much money to fix.
C. Tell the contractor and ensure they install the water line properly.
D. Fire the contractor and hire a new one who can do the job correctly.

22. Determine the resulting moment about the base of the structure due to the forces shown.

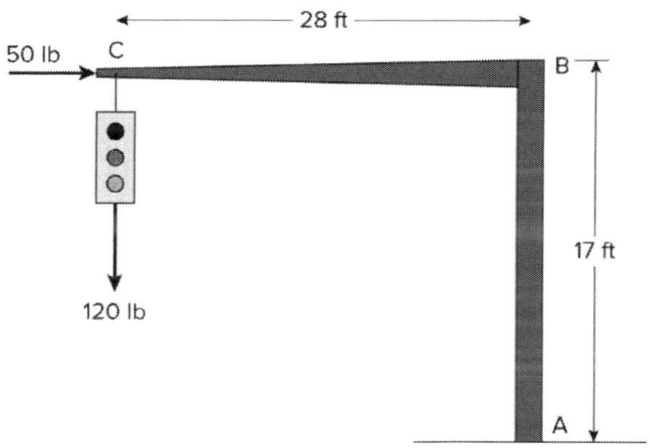

A. 610 ft·lb
B. 1,400 ft·lb
C. 2,150 ft·lb
D. 2,510 ft·lb

23. A segment of spring is clamped to the floor with a mass of 3.0-kg ball attached to the tip. A force of 6 N is required to displace the ball 25 cm horizontally. Find the spring constant and the period of the system's vibration.

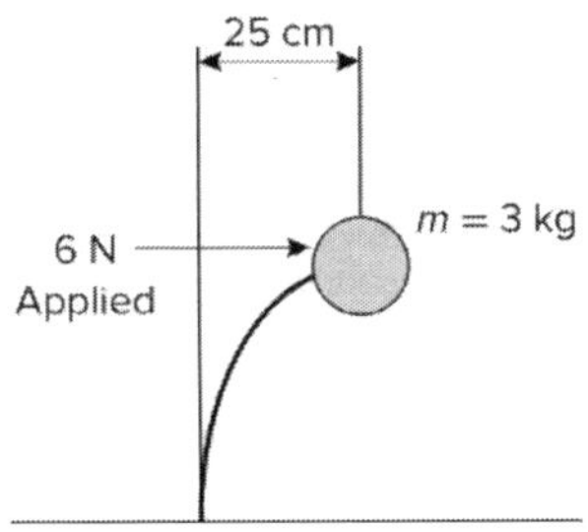

A. 12 N/m, 2.2 s
B. 12 N/m, 3.5 s
C. 24 N/m, 2.2 s
D. 24 N/m, 3.5 s

24. An 8-m rod with a cross-sectional area of 0.04 m^2 is elongated by 0.4 mm under tensile force P = 200 kN. The rod is made of which material?

A. Aluminum (E = 110 GPa)
B. Steel (E = 200 GPa)
C. Cast iron (E = 100 GPa)
D. Wood (E = 11 GPa)

25. Review the figures below and mark each one that represents a concurrent force system.

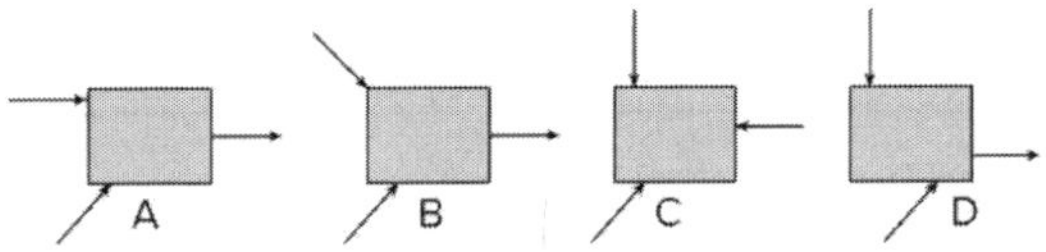

26. Determine the current (I_3) in the circuit below.

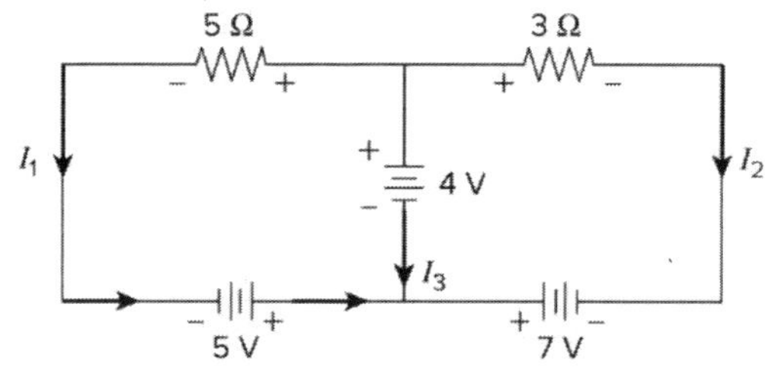

A. 1.75A ↑
B. 1.87A ↓
C. 2.2A ↓
D. 5.47A ↑

27. A beam with a circular cross section of radius 0.1 m is subject to a bending moment of 100 kN·m. Determine the normal stress on the neutral axis.

A. -66.67 MPa
B. 0 MPa
C. 33.33 MPa
D. 66.67 MPa

28. Determine the acceleration of an object with a mass of 100 lbm being acted on by a 100-lbf force in the horizontal (x) direction. Assume all friction and losses are negligible.

A. 0.03 ft/s^2
B. 16.2 ft/s^2
C. 32.2 ft/s^2
D. 100 ft/s^2

29. A random variable (T) follows a t-distribution with n degrees of freedom (n is an integer greater than 0). Determine the probability that T is greater than 0, $P(T > 0)$.

A. 0.25
B. 0.50
C. 0.75
D. 1.00

30. An engineer is a client's representative when accepting bids for a new office construction project. The engineer will determine the lowest bid and award the contract to that company. One of the sealed bids submitted has two concert tickets and a gift card to a restaurant. This same bid is determined as the lowest received. What should the engineer do in this situation?

A. Keep the gifts and accept the bid.
B. Return the gifts and accept the bid.
C. Keep the gifts and reject the bid.
D. Return the gifts and reject the bid.

31. Find the following integral using the trapezoidal rule ($n = 1$).

$\int_0^{0.6} (\cos(x) + x)dx$

A. 0.452
B. 0.563
C. 0.777
D. 0.985

32. A parallel plate capacitor has a capacitance of 6 μF with air as the dielectric. Determine the capacitance when a dielectric constant of 4 is used.

A. 12 μF
B. 14 μF
C. 18 μF
D. 24 μF

33. A force of 60 kips is applied at a point (p) in space. The distance to the point (r) from the origin is 12 ft. The force acts at an angle 55° from parallel of the distance to the point. Determine the resultant moment about the origin.

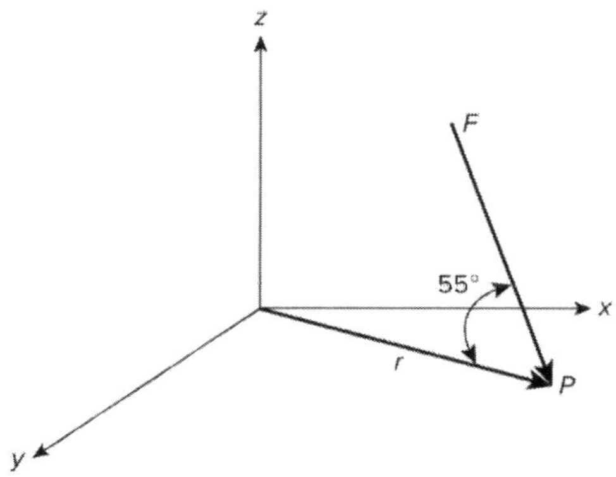

A. 440 kip · ft
B. 590 kip · ft
C. 720 kip · ft
D. 830 kip · ft

34. A spring is hanging on an inclined plane as shown in the figure below. The spring is compressed by 0.5 m and the mass is released and stops 1 m from the spring's uncompressed position on the inclined plane. What is the coefficient of friction between the surface and the block?

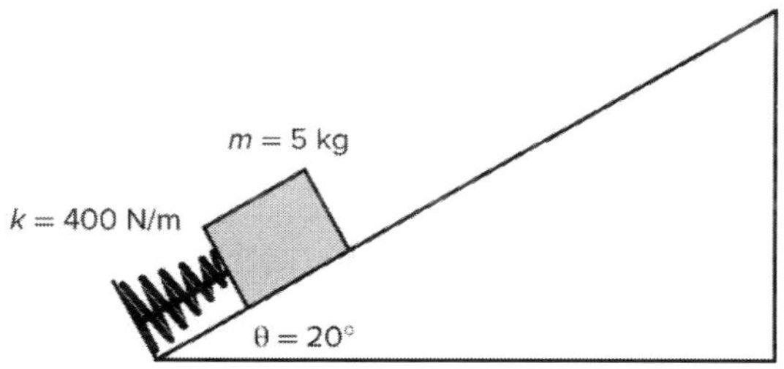

A. 0.20
B. 0.25
C. 0.36
D. 0.54

35. There are two bars made with the same material, having the same cross sections and supporting the same tension forces. The elongation in bar A is three times the elongation in bar B. Which of the following statements is correct?

A. The strain in bar A is three times the strain in bar B.
B. The strain in bar A is one-third the strain in bar B.
C. The length of bar A is three times the length of bar B.
D. The length of bar A is one-third the length of bar B.

36. Determine the force in member CD in the plane truss shown below. Joint C is pinned, and joint A can slide vertically.

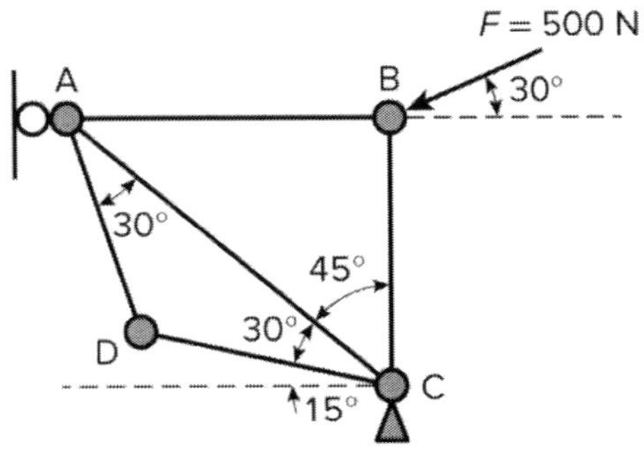

A. 0
B. 250 N (compression)
C. 500 N (compression)
D. 500 N (tension)

37. As the load is applied, edge AB moves 0.03 mm to the right as shown in the figure below. What is the most accurate shear modulus, in MPa?

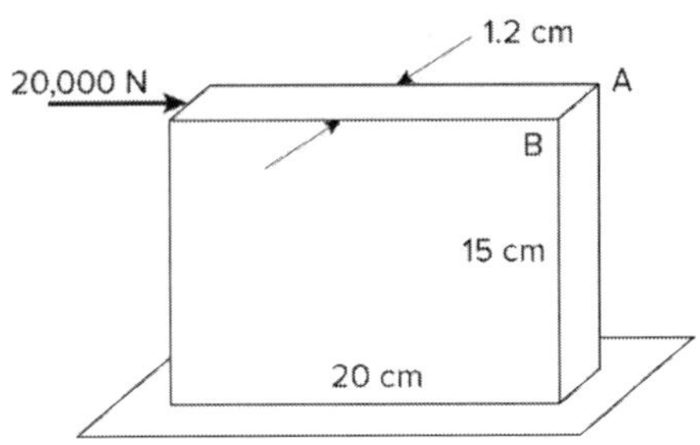

A. 32,550.10 MPa
B. 38,577.20 MPa
C. 41,666.50 MPa
D. 50,288.28 MPa

38. What is the spring constant of a mass-spring system that has a mass of 0.50 kg and oscillates with a period of 0.2 s?

A. 100 N/m
B. 425.5 N/m
C. 450 N/m
D. 493.5 N/m

39. A three-pole AC motor has a speed of 1,800 rpm. What is the period of the motor?

A. 0.212 s
B. 0.754 s
C. 3.563 s
D. 4.712 s

40. Determine the magnitude of the resultant force for a vector with force components of $35i + 65j$ kips.

A. 44 kips
B. 53 kips
C. 65 kips
D. 74 kips

41. A 6-m prismatic bar has a circular cross section with a diameter of 0.02 m. After being subjected to compressive forces at its ends, the bar's length and diameter are $L' = 5.946$ m and $D' = 0.020045$ m. What is the Poisson's ratio of the material?

A. 0.20
B. 0.25
C. 0.30
D. 0.35

42. A ball with a mass of 0.2 lbm and a velocity of 1,000 ft/s makes a perfectly elastic collision perfectly normal to the surface of a rigid object. If the initial and final velocity of the rigid object is 0, then the following is true:

A. All the kinetic energy from the ball is transferred to the object, and the ball will come to rest at the collision site.
B. None of the kinetic energy of the ball is transferred to the object, and the ball will rebound.
C. All the kinetic energy from the ball is transferred to the object, and the ball will rebound.
D. None of the kinetic energy from the ball is transferred to the object, and the ball will come to rest at the collision site.

43. What is the distance between A(4,7,8) and B(3,8,5)?

A. 2.7
B. 3.3
C. 4.3
D. 5.5

44. If you make monthly payments of $300 for the next ten years with an interest rate of 4% (compounded annually), what is the present worth of the payments?

A. $2,567
B. $3,602
C. $29,735
D. $36,000

45. The radius of gyration about the *x* axis of the area shown is:

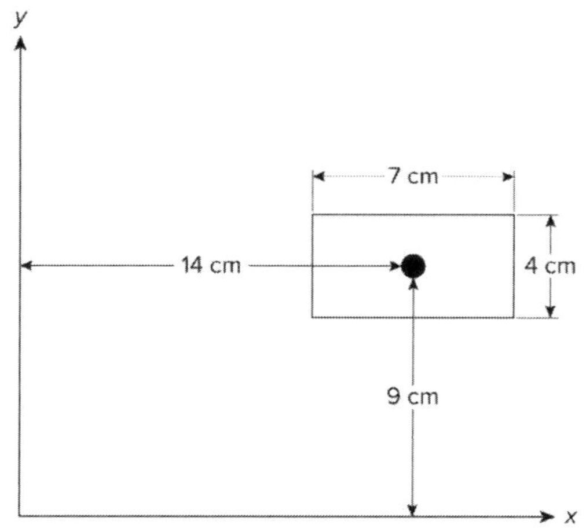

A. 6.76 cm
B. 7.09 cm
C. 9.07 cm
D. 11.7 cm

46. A cantilevered beam is subject to an inclined distributed load as shown below. Which of the following statements is not correct?

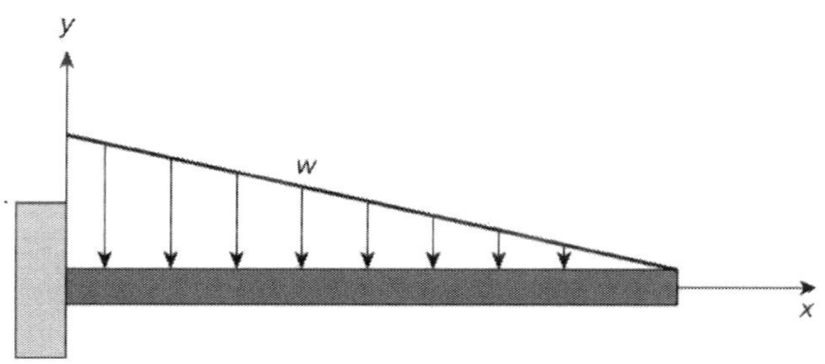

A. The deflection at the free end is maximum.
B. The slope at the free end is maximum, and it is positive.
C. The slope at the free end is maximum, and it is negative.
D. Both A and C

47. What is the natural frequency of a vertically oriented spring mass system if δ_{st} is the static deflection?

A. $\omega_n = \sqrt{\frac{g}{\delta_{st}}}$

B. $\omega_n = \sqrt{\frac{mg}{\delta_{st}}}$

C. $\omega_n = \sqrt{\frac{m}{\delta_{st}}}$

D. None of the above

48. A load of 35 kN is placed on a 3-m-long bar with a 1-cm^2 cross-sectional area. How much will the length decrease? E = 70 GPa

A. 0.15 mm
B. 0.75 mm
C. 9.0 mm
D. 15 mm

49. Determine the force in member BD in the plane truss shown below. Joints A and C are pinned.

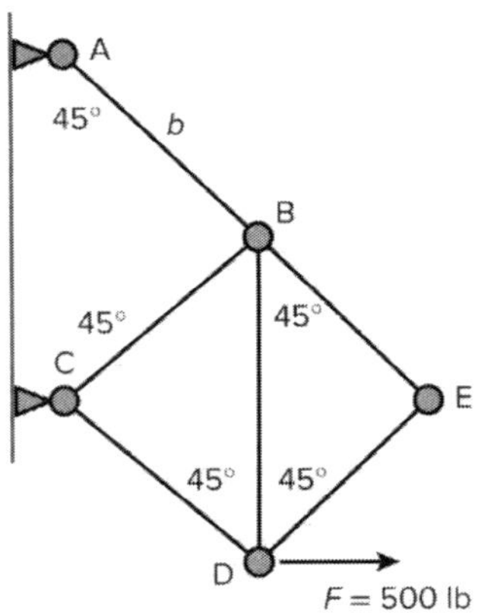

A. 500 lb (compression)
B. 500 lb (tension)
C. 707 lb (tension)
D. 1,000 lb (compression)

50. A restoring force is always one that: (Select as many as apply.)

A. is opposite in direction to displacement.
B. is in the same direction as displacement.
C. can be either in the same direction or in the opposite direction of displacement.
D. obeys Hooke's law.
E. does not obey Hooke's law.

51. A system of pulleys fixed to a ceiling holds up a block that weighs 150 kips. Determine the force (T) that would be required to keep this system static. Assume friction can be neglected.

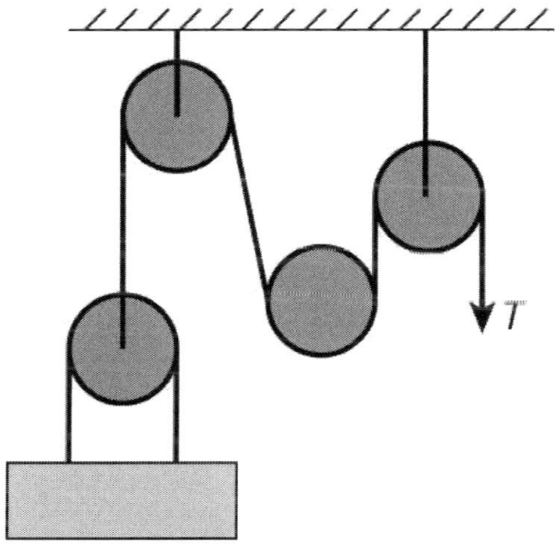

A. 37.5 kips
B. 50 kips
C. 75 kips
D. 150 kips

52. A projectile is launched from a cannon at 2,500 ft/s at an angle of 40° from the ground. Eric is observing from 250 ft behind where the cannons are stationed. How long will it take for the projectile to be a total of 2,500 ft from Eric?

A. 1.12 s
B. 1.17 s
C. 1.55 s
D. 2.00 s

53. In a conic section, the eccentricity is $e = 1.76$. This conic section most probably is a(n) ____________.

54. A flywheel rotating at 650 rpm is quickly brought to a halt in 0.5 s. How many revolutions does the flywheel travel before it is brought to a complete stop?

A. 1.4 rev
B. 2.7 rev
C. 3.5 rev
D. 6.1 rev

55. Which of the following is a sign of brittle failure?

A. Plastic deformation
B. Cleaved fracture surface
C. Fibrous fracture surface
D. Elastic deformation

56. A material with an initial cross-section area of 0.52 in^2 has been tensile tested to a load of 62,255 lbf with an engineering strain of 0.0041 in/in. What is most likely the modulus of elasticity of the material in psi?

A. 491 psi
B. 7,900,000 psi
C. 29,200,000 psi
D. 29,300,000 psi

57. The barometer shown is at 20°C and a pressure of 1 atm. The vapor pressure of the unknown fluid at 20°C is 2.3 kPa. Determine the maximum height for the fluid in the barometer. The specific weight of fluid = 8,703 N/m^3.

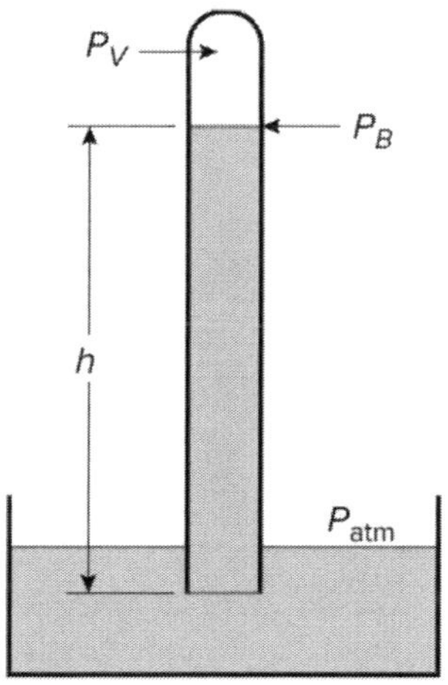

A. 9.6 m
B. 11.4 m
C. 13.8 m
D. 14.1 m

58. The degree of freedom of a two-component system (water and alcohol) that contains three phases is _______________.

59. Which one of the following is the correct expression of Fourier's law of conduction? Assume that Q = rate of heat transfer, k = thermal conductivity, A = area through which heat is being transferred, x = distance in direction of heat transfer, T = temperature, c_p = heat capacity, $a = \frac{k}{(rc_p)}$, and r = density.

A. $Q = -\left(\frac{1}{a}\right)\left(\frac{dT}{dx}\right)$
B. $Q = -c_pA\left(\frac{dT}{dx}\right)$
C. $Q = -aA\left(\frac{dT}{dx}\right)$
D. $Q = -kA\left(\frac{dT}{dx}\right)$

60. What is the damped natural frequency of a second-order control system where the natural frequency is 32 Hz and the damping ratio is 0.76?

A. 0
B. 3.2
C. 12.6
D. 20.8

61. A member has tensile strength of 50 kpsi and compressive strength of 150 kpsi. If it is subject to a general stress state with principal stresses of 30 kpsi, -20 kpsi, and -40 kpsi, determine whether the material failure occurs according to the Coulomb-Mohr theory.

A. The material failure does not occur because $\sigma_1 < S_{ut}$.
B. The material failure does not occur because $\sigma_3 > -S_{ut}$.
C. The material failure does not occur because $\frac{\sigma_1}{S_{ut}} - \frac{\sigma_3}{S_{uc}} < 1$.
D. The material failure occurs because $\frac{\sigma_1}{S_{ut}} - \frac{\sigma_3}{S_{uc}} \geq 1$.

62. An engineer is designing a part to be used in an application where the part will be subjected to tensile, bending, and fatigue cycles in a moderately corrosive environment. With respect to fatigue strength, which material should be selected?

A. A1 2014-T651
B. 4340 steel
C. 52100 steel
D. A1 2024-T3

63. In a Carnot cycle, a system operates between two thermal reservoirs, T_H = 400 K and T_L = 300 K. Read the P-V diagram below. Circle the part of the process that indicates the system is experiencing an adiabatic expansion.

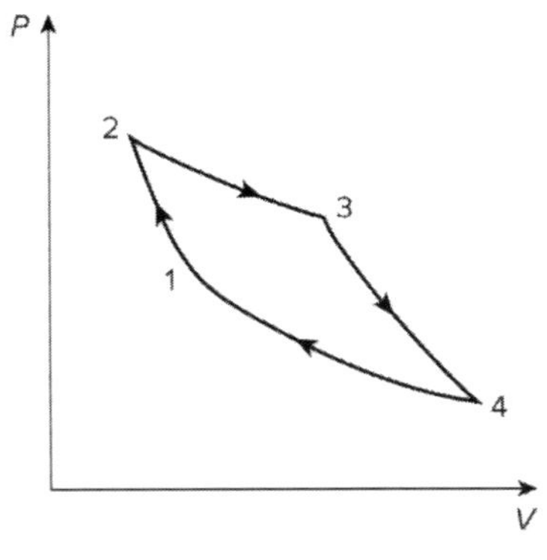

64. Calculate the voltage resolution of the following analog-to-digital signal converter. The lowest voltage is 12 V and the highest voltage is 120 V. There are four conversion bits.

A. 0.42
B. 3.75
C. 6.75
D. 15

65. A thermometer is applied to measure the body temperature of a patient. The thermometer has the following characteristics: length (l) = 10 mm, diameter (d) = 4 mm, h = 85 W/m^2-K, and k = 8.1 W/m-K. What is the Biot number (Bi)?

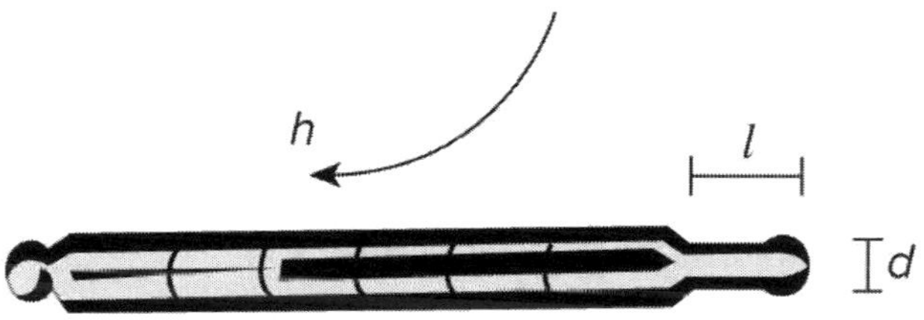

A. 1.2×10^{-3}
B. 4.3×10^{-3}
C. 8.6×10^{-3}
D. 9.5×10^{-3}

66. The water temperature in a glass tube is 20°C. Determine the diameter of glass tube that is required to keep the capillary rise of the water below 3 mm. Assume that the surface tension is 0.0728 N/m and β = 0°.

A. 5.60 mm
B. 9.92 mm
C. 10.82 mm
D. 12.80 mm

67. You are designing components that require a specific fit between a shaft and hole. This will require an interference fit for parts requiring rigidity and alignment with critical accuracy. Which of the following designations should be indicated on the part drawings to ensure the proper fit between components?

A. H7/h6
B. H8/f7
C. H11/c11
D. H7/p6

68. A gear drive with a module of 4 mm and pressure angle of 20° has 80 teeth on gear and 20 teeth on pinion. What are the circular pitch, base pitch, and length between shafts?

A. 10 mm, 11 mm, 100 mm
B. 10.5 mm, 11.5 mm, 150 mm
C. 12.6 mm, 11.8 mm, 200 mm
D. 13 mm, 12 mm, 220 mm

69. A solid sphere (diameter = 44 mm) is falling through the air. If the coefficient of drag is 0.6, determine the terminal velocity. $\left(\rho_{sphere} = 2{,}340\ \text{kg/m}^3, \rho_{air} = 1.225\ \text{kg/m}^3\right)$

A. 29.6 m/s
B. 32.4 m/s
C. 42.8 m/s
D. 58.9 m/s

70. A 4340 steel block undergoes stress of 50 ksi. What is the largest possible interior crack size that the material can possess before fracture? Assume a geometrical factor of 1.1.

A. 0.186 in
B. 0.25 in
C. 0.50 in
D. 0.625 in

71. An electric resistance heater is in a room rated at 8 kW, and it is kept running for 15 minutes. What is the energy transferred to the room by the heater?

A. 1 kJ
B. 8 kJ
C. 120 kJ
D. 7,200 kJ

72. A spring is designed for a force of 870 N over a scale of 72-mm length and is operating in a casing. The diameter of the casing is 36 mm. The number of turns is around 43 with a modulus of rigidity of 87 kN/mm^2. The spring parameters will be:

A. D = 11.91 mm, d = 3 mm
B. D = 12.00 mm, d = 3 mm
C. D = 13.23 mm, d = 2 mm
D. D = 14.43 mm, d = 5 mm

73. Calculate the percent overshoot of a second-order system with a damping ratio of 0.84 subjected to a step unit input.

A. 0.28%
B. 0.77%
C. 28%
D. 77%

74. A 1-kg block of steel at 40°C is placed in open air (atmosphere) at 10°C and allowed to cool. Which one of the following is generally true? Assume that thermal conductivity and heat capacity are not a function of temperature between 40°C and 10°C.

A. The heat capacity of the steel varies with time.
B. The thermal conductivity of the steel varies with time.
C. The thermal energy of the steel remains constant.
D. The thermal energy of the steel varies with time.

75. Water is heated by refrigerant R134a inside of a heat exchanger, where R134a is in a two-phase state with pressure (P) = 0.1 MPa and vapor quality (x) = 0.2. What is the entropy of the refrigerant?

A. 0.8 kJ/kg·K
B. 1.08 kJ/kg·K
C. 1.2 kJ/kg·K
D. 2.3 kJ/kg·K

76. A force of 31 N is required to keep the top plate at a constant velocity of 8 m/s. If a consistency index of fluid is 0.44 Pa $s^{-1.2}$, find the power law index.

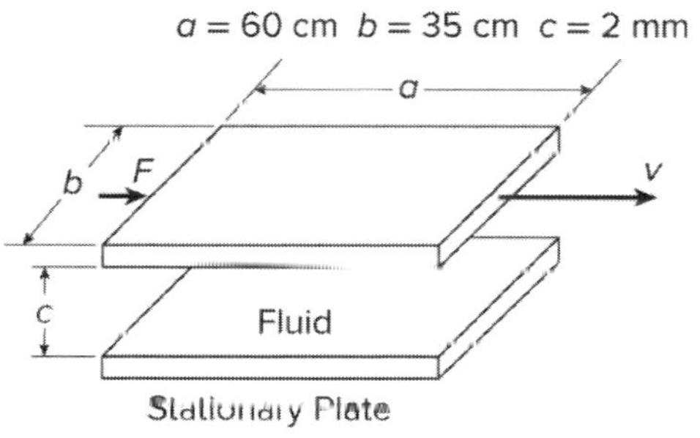

A. 0.60
B. 0.70
C. 1.80
D. 2.40

77. An aluminum alloy is heated from room temperature (70°F) to 800°F. What is the thermal strain of the aluminum alloy given that aluminum's average coefficient of linear thermal expansion is 12.8 × 10-6/°F.

A. 9.34 × 10-3
B. 10.24 × 10-3
C. 11.45 × 10-3
D. 14.60 × 10-3

78. What should the sampling rate be for a television that needs to update its display in a time period of 0.0083 s?

A. 60 Hz
B. 120 Hz
C. 600 Hz
D. 1,200 Hz

79. A member has yield strength of 60 kpsi. If it is subject to a general stress state with the principal stresses of -40 kpsi, -20 kpsi, and 30 kpsi, determine if material failure occurs according to the distortion-energy theory.

A. The material failure occurs because $\sigma' > S_y$.
B. The material failure does not occur because $\sigma' < S_y$.
C. The material failure does not occur because $\sigma_1 < S_y$.
D. The material failure occurs because $\sigma_3 < -S_y$.

80. Water at 30°C is flowing through an open channel with a volumetric flow rate of 1 m³/s. The height of the channel is 1.7 m, and the width is 3 m. What is the Reynolds number?

A. 2.85 × 10⁵
B. 4.29 × 10⁵
C. 7.84 × 10⁵
D. 8.28 × 10⁵

81. During the power stroke of the cylinder in an Otto cycle, the air expands from an initial pressure to a final pressure of 16 bars. What is the initial pressure if the thermal efficiency of the cycle is 50%?

A. 265 kPa
B. 283 kPa
C. 289 kPa
D. 293 kPa

82. Which of the following could most easily be used to determine the weight of a suitcase?

A. Strain gauge
B. PID controller
C. Thermocouple
D. Optical sensor

83. The 16-gauge stainless steel plate shown here is going to be drilled to create two holes. All measurements are in inches. What is the maximum possible distance between the inner edges of the holes?

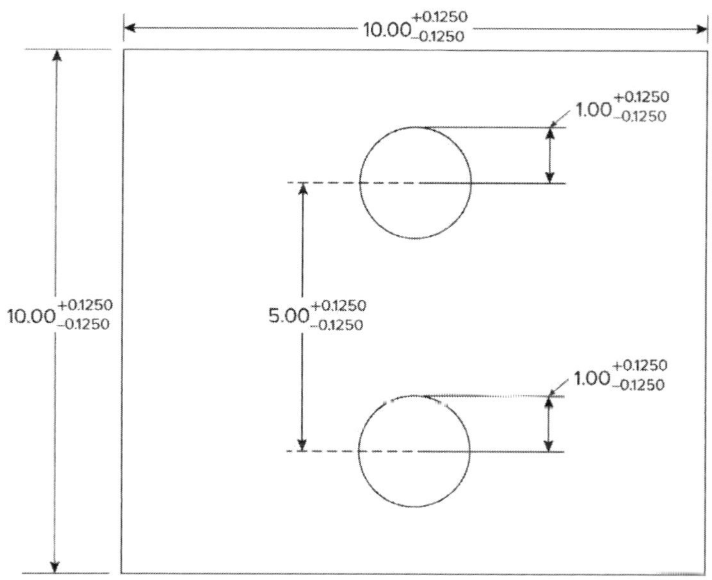

A. 2.625 in
B. 3.375 in
C. 4.750 in
D. 5.000 in

84. A furnace wall is made of 250-mm firebrick with a conductivity of 1.07 W/m-K. If the area of the wall is 1.2 m^2, what is the plane wall conduction resistance?

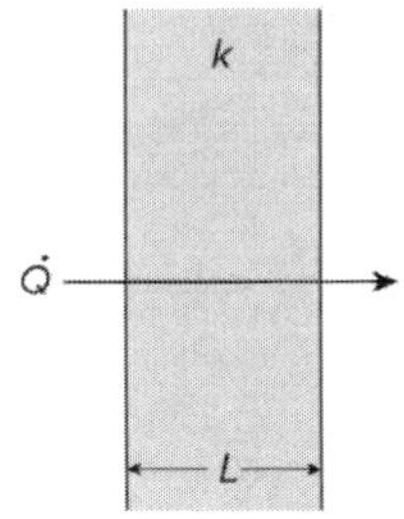

A. 0.195 K/W
B. 0.538 K/W
C. 0.918 K/W
D. 1.212 K/W

85. When steel is cold worked, the grains do which of the following?

A. Dislocate
B. Recrystallize
C. Compress
D. Nucleate

86. In the figure below, the flow rate in pipe 1 is 800 gpm and the flow rate in pipe 2 is 500 gpm. What is the flow rate in pipe 3?

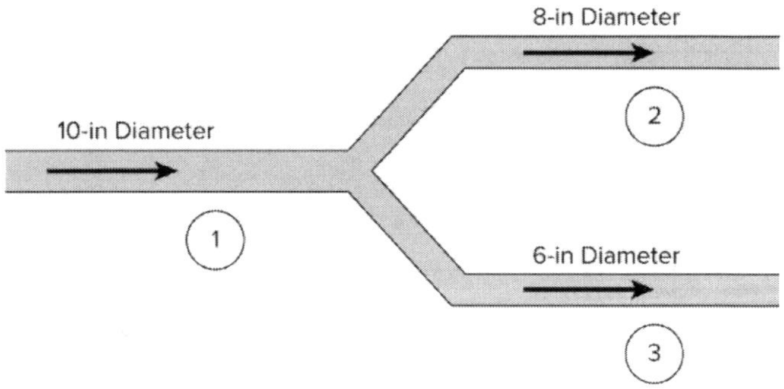

A. 100 gpm
B. 200 gpm
C. 300 gpm
D. 400 gpm

87. If the temperature of ice changes from 273 to 215 K, what is the change in enthalpy? (c_p = 2.05 kJ/kg·K)

A. -235 kJ
B. -119 kJ
C. 119 kJ
D. 235 kJ

88. A copper pipe has the following characteristics: length (L) = 2 m, inner radius (r_1) = 2 cm, outside radius (r_2) = 2.5 cm, outside surface temperature (T_2) = 45°C, inner surface temperature (T_1) = 15°C, and copper thermal conductivity (k) = 285 W/m-K. What is the conduction through the cylindrical wall?

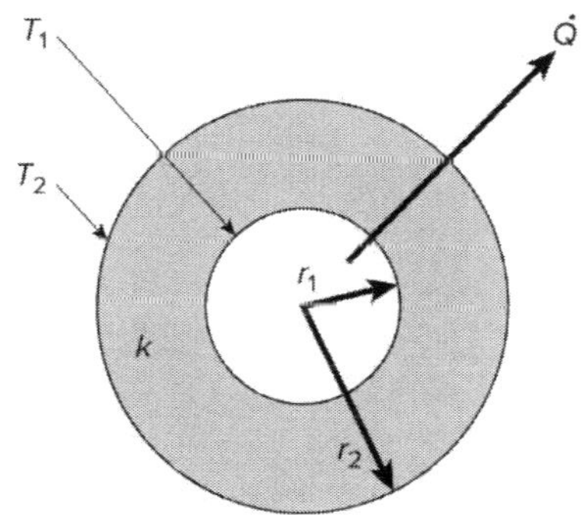

A. 350.2 kW
B. 481.3 kW
C. 700.5 kW
D. 800.4 kW

89. A member has tensile strength of 60 kpsi and compressive strength of 100 kpsi. According to the maximum-normal-stress theory, under which plane stress state will material failure not occur?

A. -50 kpsi, -180 kpsi
B. -60 kpsi, -130 kpsi
C. -70 kpsi, -90 kpsi
D. All of the above

90. Which of the following is not a desirable property to have in a material used in the construction of a pressure vessel?

A. High yield strength
B. High melting temperature
C. High resistance to corrosion
D. High ductile-to-brittle transition temperature

91. Find the heat transfer per meter of length for the system shown below. Assume that T_i = 100°C, T_o = 40°C, r_1 = 3.5 cm, r_2 = 5 cm, r_3 = 7 cm, k_A = 15 W/m°C, and k_B = 0.08 W/m°C.

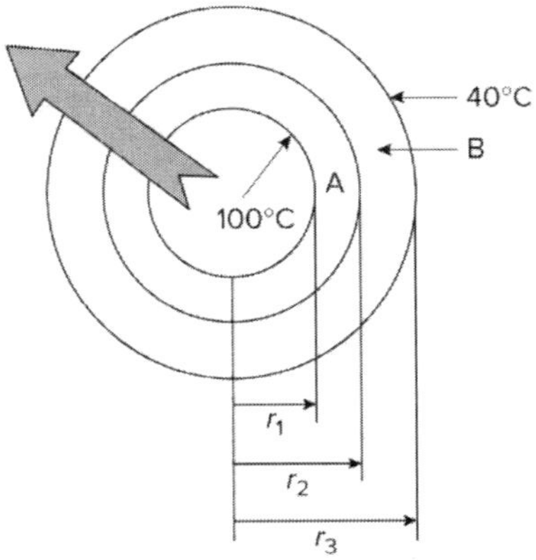

A. 64 W/m
B. 76 W/m
C. 89 W/m
D. 98 W/m

92. A 1/2-in steel bolt clamping steel with a bolt grip of 2 in has a joint coefficient of 0.168 and is preloaded with 70 lbf. If a 38-lbf external force is applied on the bolt, what is the load carried by the bolt?

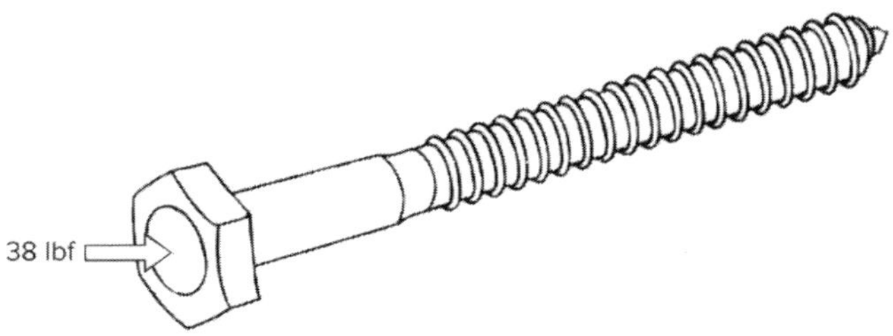

A. -76.38 lbf
B. 76.38 lbf
C. 76.38 kPa
D. 76.38 N

93. Carbon monoxide is mixed in air with volume fraction of 0.04 in a vessel containing water. Air pressure = 1 atm. What is the partial pressure of carbon monoxide?

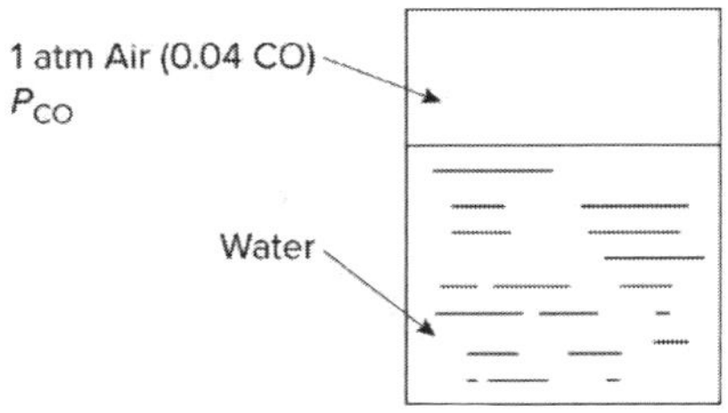

A. 0.04 atm
B. 0.50 atm
C. 0.96 atm
D. 1.00 atm

94. A 24-in diameter concrete pipe (C = 120) is 2,000 ft long and has a head loss of 10.2 ft. Determine the discharge capacity of the pipe according to the Hazen-Williams formula. Assume the hydraulic radius is 0.75.

A. 17.3 ft^3/s
B. 18.4 ft^3/s
C. 21.9 ft^3/s
D. 24.0 ft^3/s

95. What type of failure is caused by repeated loading?

A. Fatigue failure
B. Ductile failure
C. Brittle failure
D. Void failure

96. Hot and cold fluids have heat transfer in a parallel flow concentric tube heat exchanger as shown in the following image. If $NTU = 1.1$ and $C_r = 0.8$, what is the heat exchanger effectiveness (ε)?

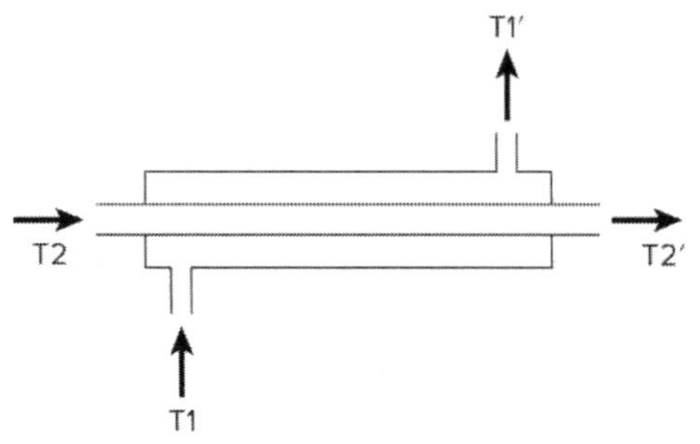

A. 0.28
B. 0.32
C. 0.48
D. 0.62

97. A 1/2-in steel bolt clamping steel with a bolt grip of 3 in has a total bolt load of 80 lbf. If the external load applied is 50 lbf, what is the load carried by the members?

The table below shows computation of bolt and member stiffness. The steel members are clamped using a 1/2-in 13 NC steel bolt.

	Stiffnesses, *M* (lbf/in)	
Bolt Grip (in)	k_b	k_m
2	2.57	12.69
3	1.79	11.33
4	1.37	10.63

A. -30 lbf
B. 30 lbf
C. 30 N
D. 30 N-mm

98. A Carnot engine operating between 50°C and 2,500°C is modified solely by raising the high temperature by 250°C and raising the low temperature by 80°C. Which of the following statements about the engine is true?

A. The thermodynamic efficiency increases.
B. The thermodynamic efficiency decreases.
C. The thermodynamic efficiency increases by 7%.
D. None of above

99. Assuming atmospheric pressure of 101.3 kPa, determine the absolute pressure at point *a* in the figure, at the center of the gasoline tank. (SG_{Hg} = 13.6, SG_{gasoline} = 0.91)

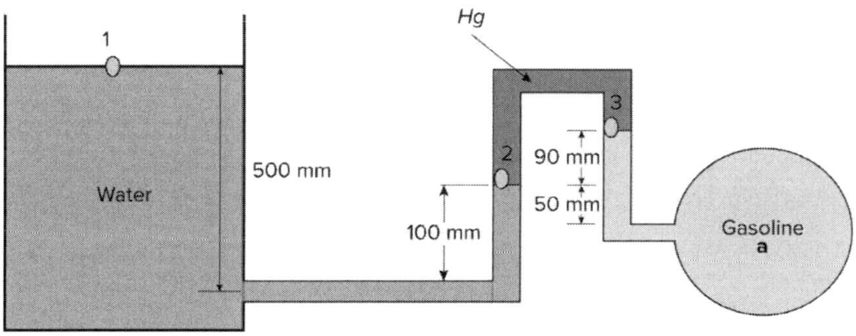

A. 87.1 kPa
B. 94.47 kPa
C. 109.3 kPa
D. 128.7 kPa

100. A composite consists of continuous glass fiber that is reinforced with epoxy resin. The glass fiber has a tensile strength of 375,000 psi and occupies 72% by volume. The epoxy resin has a tensile strength of 8,000 psi. What is most nearly the tensile strength of the composite in psi?

A. 110,050 psi
B. 200,240 psi
C. 272,250 psi
D. 273,250 psi

101. A mechanism consists of a spring for tension capacity with a load of 45 N. The wire diameter of the spring is 5 mm, with a spring index of 8 and 20 active coils. The design shear stress is 720 MPa and the modulus of rigidity is 60 kN/mm^2. The force required by the spring to reach its yield strength is:

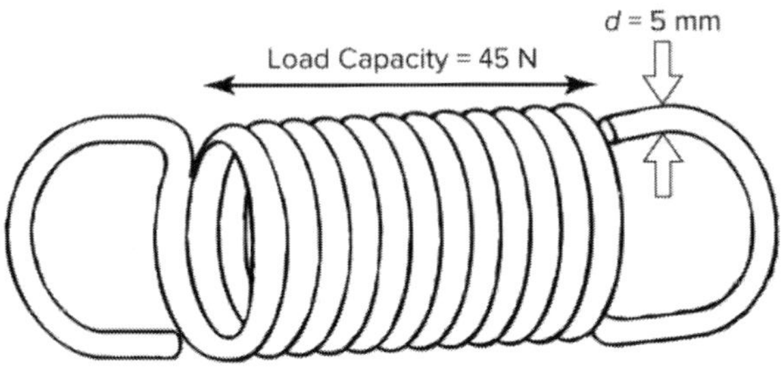

A. 760.8 N
B. 821.0 N
C. 831.6 N
D. 890.7 N

102. A 9-cm plane wall is exposed to 39°C air flow on one side and 11°C air flow on the other side. What are the surface temperatures of the warmer side and cooler side if the convection coefficients of both fluids are 22 W/(m^{2}K)? The thermal conductivity of the wall is 13 W/(mK).

A. 25°C; 25°C
B. 26°C; 24°C
C. 28°C; 22°C
D. 32°C; 18°C

103. In a counter-flow, double-pipe heat exchanger, the mass flow rate of a Freon refrigerant is 0.8 kg/s and the enthalpy at the inlet and outlet is 310 kJ/kg and 85.4 kJ/kg, respectively. What is the rise in temperature of water (coolant) if the mass flow rate of water is 1.4 kg/s?

A. 15.25°C
B. 22.3°C
C. 30.7°C
D. 35.0°C

104. Determine the friction factor for flow in a circular pipe with a Reynolds number of 984.

A. 0.028
B. 0.084
C. 0.041
D. 0.065

105. The stress applied to the metal body shown below comes to 200 MPa. What is the smallest critical K_{IC} value that would prevent the crack from propagating throughout the material?

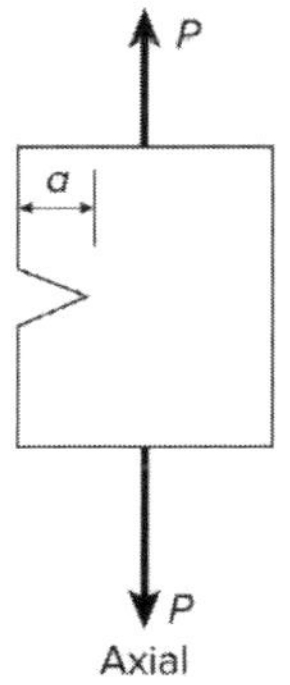

a = 5 mm
σ = 200 MPa

A. 22 MPa
B. 26 MPa
C. 30 MPa
D. 34 MPa

106. If the input velocity of a two-set gear train is 24 rad/s and velocity ratio is 0.5, what is the output velocity achieved by the gear train?

A. 10 rad/s
B. 12 rad/s
C. 14 rad/s
D. 16 rad/s

107. If atmospheric conditions are 101.3 kPa, 20°C, and specific humidity of 0.0095 kg/kg of dry air, what is the dew point temperature?

A. 5°C
B. 9°C
C. 13°C
D. 17°C

108. Water flows through a Borda orifice. The cross-sectional area of the entrance is 25 in^2. The pressure difference is 10 lb/ft^2 and there is no change in elevation. Determine the flow rate (Q) in ft^3/s.

A. 0.28 ft^3/s
B. 2.2 ft^3/s
C. 3.4 ft^3/s
D. 41.0 ft^3/s

109. Refrigerant R134a has a mass flow rate of 15 kg/s when entering the compressor at 0.06 MPa and 40°C. The compression process is isentropic, and refrigerant leaves the compressor at 0.4 MPa. Then the refrigerant is cooled to a saturated liquid in a condenser in an isobaric process. After that, the refrigerant experiences a throttling process in an expansion valve before entering the evaporator at 0.06 MPa. Assume the enthalpy remains the same in the expansion valve. What is the coefficient of performance (COP) of this refrigeration cycle?

A. 3.1
B. 4.4
C. 6.2
D. 7.9

110. A glass tube is 30 cm long. The bottom 15 cm contains water. The top 10 cm contains gas (specific gravity = 0.6) and the remaining 5 cm is air. Determine the pressure at the bottom of the tube.

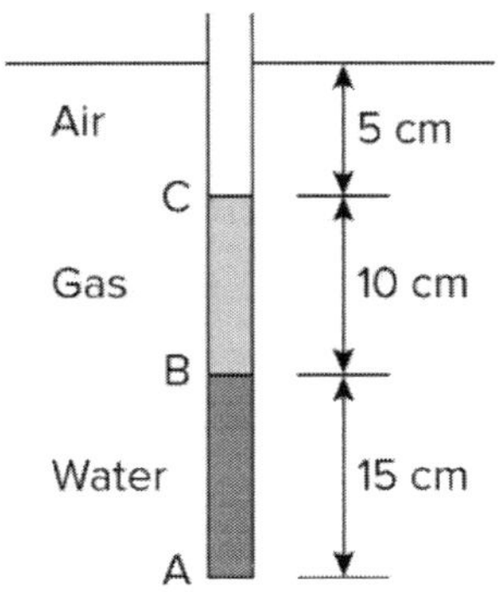

A. 142 Pa
B. 1,980 Pa
C. 2,060 Pa
D. 3,956 Pa

SOLUTIONS

QUICK SOLUTIONS REFERENCE

Question #	Answer	Question #	Answer	Question #	Answer
1.	See Solution	40.	D	79.	A
2.	A	41.	B	80.	C
3.	C	42.	B	81.	B
4.	A	43.	B	82.	A
5.	A	44.	C	83.	B
6.	A, B	45.	C	84.	A
7.	B	46.	B	85.	A
8.	A, B, D	47.	A	86.	C
9.	B	48.	D	87.	B
10.	A	49.	A	88.	B
11.	D	50.	A, D	89.	C
12.	D	51.	D	90.	D
13.	C	52.	B	91.	C
14.	A, B, D	53.	See Solution	92.	B
15.	C	54.	B	93.	A
16.	B	55.	B	94.	D
17.	C	56.	C	95.	A
18.	B	57.	B	96.	C
19.	A, B	58.	See Solution	97.	A
20.	A	59.	D	98.	B
21.	C	60.	D	99.	B
22.	D	61.	C	100.	C
23.	C	62.	B	101.	C
24.	C	63.	See Solution	102.	B
25.	B	64.	C	103.	C
26.	D	65.	D	104.	D
27.	B	66.	B	105.	C
28.	A	67.	D	106.	B
29.	B	68.	C	107.	C
30.	B	69.	C	108.	A
31.	C	70.	A	109.	B
32.	D	71.	D	110.	C
33.	B	72.	A		
34.	C	73.	B		
35.	C	74.	D		
36.	A	75.	B		
37.	C	76.	B		
38.	D	77.	A		
39.	A	78.	B		

1. $\frac{\text{Change in } Y}{\text{Change in } X} = \frac{Y_C - Y_A}{X_C - X_A} = \frac{4-0}{0-(3)} = \frac{4}{-3}$

To find the slope, stand at A and look at the vertical axis, counting four spaces up the line (for the slope of 4/-3). The change in Y is $(Y_C - Y_A)$, and the change in X is $(X_C - X_A)$. The mark should appear at the fourth hash mark on the vertical axis.

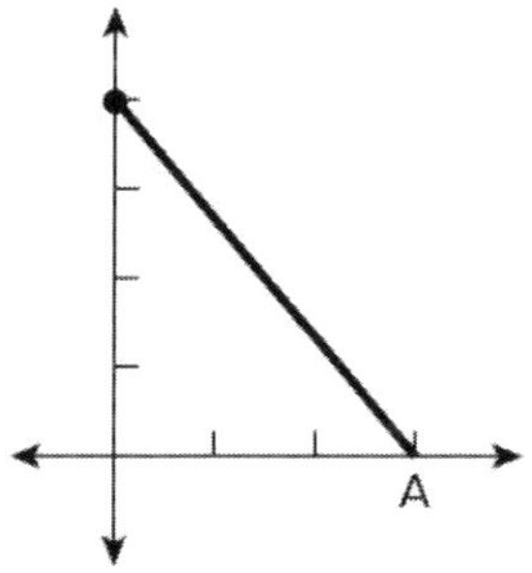

Reference: NCEES *FE Reference Handbook* > Mathematics > Straight Line

2. On a Casio FX-115ES calculator, do the following.
1. MODE
2. 3 – FOR STAT
3. 2 FOR A+BX
4. INPUT X AND Y NUMBERS (followed by =)
5. PRESS ON
6. SHIFT (followed by 1)
7. 5 for Reg
8. 3 for R correlation
9. EQUAL

Alternate method:

$$R = \frac{n(\sum xy) - (\sum x)(\sum y)}{\sqrt{[n\sum x^2 - (\sum x)^2]\,[n\sum y^2 - (\sum y)^2]}}$$

Reference: NCEES *FE Reference Handbook* > Engineering Probability and Statistics > Sample Correlation Coefficient (R) and Coefficient of Determination (R^2)
Answer: A

3. Convert each payment to an A-pattern for the m preceding years.

$A = 100(A/F, 10\%, 2)$

$= 100(0.4762)$

$= \$47.62$

Convert the A-pattern to a P-pattern:

$P = 47.62(P/A, 10\%, 6)$

$= 47.62(4.3553)$

$= \$207$

Reference: NCEES *FE Reference Handbook* > Engineering Economics > Factor Name table, Interest Rate Tables
Answer: C

4. Draw a free-body diagram of the block on the plank, orienting the *x*-axis along the incline.

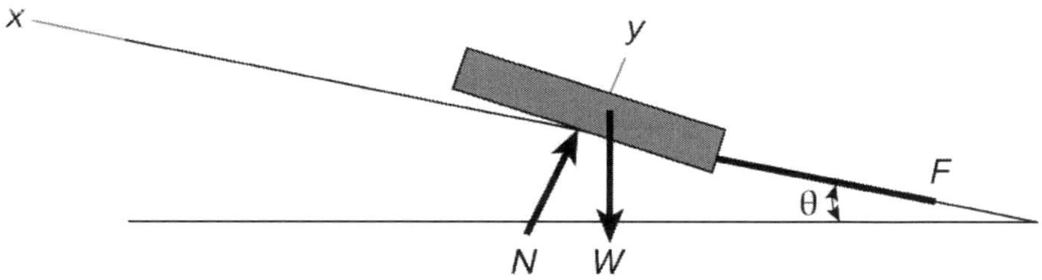

Sum the forces in the *x*-direction.

$$\sum F_x = 0 = W\sin\theta - F = 0$$

Sum the forces in the *y*-direction.

$$\sum F_y = 0 = N - W\cos\theta = 0$$

Note the equation for the force of static friction.

$$F_s = \mu_s N$$

Solve the *x* force equation for F_s.

$$F_s = W\sin\theta$$

Solve the *y* force equation for *N*.

$$N = W\cos\theta$$

Note: In the above equation, the weight of the block cancels out, leaving:

$$\tan\theta = 0.35$$

$$\theta = \tan^{-1}0.35$$

$$\theta = 19.3°$$

Reference: NCEES *FE Reference Handbook* > Statics > Resolution of a Force, Equilibrium Requirements
Answer: A

5. The element stresses are as below (tensile is positive, and compressive is negative):

$\sigma_x = -80 \text{ MPa}, \sigma_y = 50 \text{ MPa}, T_{xy} = -25 \text{ MPa}$

The center of a Mohr's circle is always located at C, which can be calculated as follows:

$$C = \frac{\sigma_x + \sigma_y}{2} = \frac{-80 \text{ MPa} + 50 \text{ MPa}}{2} = -15 \text{ MPa}$$

Reference: NCEES *FE Rcference Hundbook* > Mechanics of Materials > Mohr's Circle—Stress, 2D

Answer: A

6. Damped and undamped are the two subcategories of natural and forced vibrations.

Reference: NCEES *FE Reference Handbook* > Dynamics > Free and Forced Vibration

Answer: A, B

7. Write Kirchhoff's voltage law for each loop and simplify the middle resistor group. The 4-Ω resistors in parallel are equivalent to a 2-Ω resistor.

$$R_{eq} = \frac{1}{\frac{1}{R_1} + \frac{1}{R_2}} = \frac{1}{\frac{1}{4} + \frac{1}{4}} = 2\ \Omega$$

Resistors in series are additive and the middle branch is equivalent to a single 5-Ω resistor.

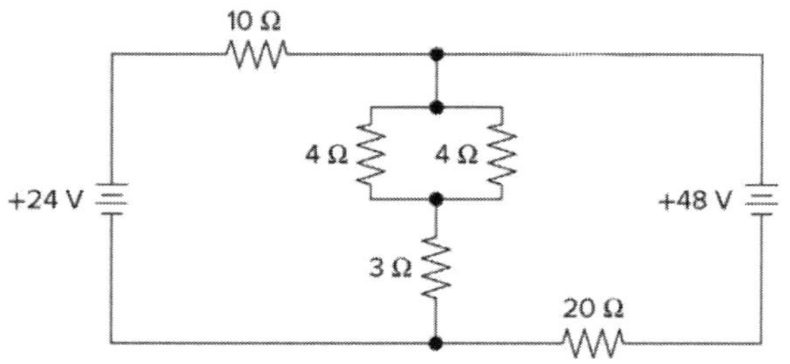

$$24 - 10I_1 - (I_1 + I_2)(5) = 0$$
$$48 - (I_1 + I_2)(5) - 20I_2 = 0$$

Solve for I_1 from the first equation.

$$\frac{24 - 5I_2}{15} = I_1$$

Substitute into the second equation.

$$48 - \left(\frac{24 - 5I_2}{15} + I_2\right)(5) - 20I_2 = 0$$

$$48 - \left(8 - \frac{5}{3}I_2 + 5I_2\right) - 20I_2 = 0$$

$$40 + \frac{5}{3}I_2 - 25I_2 = 0$$

$$I_2 = 1.71\text{A}$$

Solving for I_1:

$$I_1 = \frac{24 - 5(1.71\text{A})}{15} = 1.03\text{A}$$

Solve for the voltage across the 10-Ω resistor.

$$V = IR = (1.03\text{A})(10\ \Omega) = 10.3\ \text{V}$$

Reference: NCEES *FE Reference Handbook* > Electrical and Computer Engineering > Resistors in Series and Parallel, Kirchhoff's Laws

Answer: B

8. All of the given answers, except answer C, are fundamental rules in the NCEES Rules of Professional Conduct. The Rules state that "licensees shall issue no statements, criticisms, or arguments on engineering and surveying matters that are inspired or paid for by interested parties unless they explicitly identify the interested parties on whose behalf they are speaking and reveal any interest they have in the matters."

Reference: NCEES *FE Reference Handbook* > Ethics and Professional Practice > *Model Rules*, Section 240.15 Rules of Professional Conduct
Answer: A, B, D

9. Based on the given information, the probability that a light bulb is defective is p = 5%, and the probability that a light bulb is not defective is $q = 1 - p = 1 - 5\% = 95\%$.

Let X be the event that X light bulbs are defective in a package of n = 8 light bulbs.

Then, X follows a binomial distribution with parameters n = 8 and p = 5%.

Following the probability formula of a binomial distribution, the probability that x = 1 is defective in n = 8 light bulbs is:

$$P(x = 1) = P_8(1) = C(8,1)(5\%)^1(95\%)^{8-1} = \frac{8!}{1!\,(8-1)!}(5\%)^1(95\%)^{8-1}$$

$$\cong 0.279$$

Reference: NCEES *FE Reference Handbook* > Engineering Probability and Statistics > Binomial Distribution
Answer: B

10. For the slope-intercept form of a line:

$y = mx + b$.

b is the y intercept where $x = 0$.

m is the slope (dy/dx).

$-b/m$ is the x intercept, where $y = 0$.

$-2/7$ is the x intercept.

Reference: NCEES *FE Reference Handbook* > Mathematics > Straight Line
Answer: A

11. Assume $i = 2\%$ and n = 1 year.

Substitute values.

$(A/F, 2\%, 1) + 2\% = (A/P, 2\%, 1)$

$(1.0000) + 0.02 = 1.0200 = (A/P, 2\%, 1)$

Reference: NCEES *FE Reference Handbook* > Engineering Economics > Factor Name table
Answer: D

12. To stamp a drawing, the engineer must be competent in that specific field.

Answer A: They cannot stamp mechanical drawings because their license is in civil engineering.
Answer B: They cannot stamp electrical drawings because their license is in civil engineering.
Answer C: They cannot stamp a drawing that they did not themselves work on or oversee the work done.

Reference: NCEES *FE Reference Handbook* > Ethics and Professional Practice > *Model Rules*, Section 240.15 Rules of Professional Conduct
Answer: D

13. $K_T = K_E$ when consistent SI units are used [N·m/A and V/rad/s]

Apply the DC motor equation:

$T = K_T I = (1.3505\ \text{V/rad/s})40\text{A} = 54\ \text{N} \cdot \text{m}$

Reference: NCEES *FE Reference Handbook* > Electrical and Computer Engineering > Servomotors and Generators
Answer: C

14. Answer A is correct: The slope at the fixed end is zero.
Answer B is correct: The slope at the free end is the maximum, and it is positive because the angle is counterclockwise.
Answer C is not correct: The slope at the free end is the maximum, and it is positive.
Answer D is correct: The slope at the midpoint is negative.
Answer E is not correct.

Reference: NCEES *FE Reference Handbook* > Mechanics of Materials > Cantilevered Beam Slopes and Deflections
Answer: A, B, D

15. Since the forcing function dies too soon, the steady-state solution, which has a direct dependence on the forcing function, will also die. Therefore, the overall response becomes a transient one.

Reference: NCEES *FE Reference Handbook* > Instrumentation, Measurement, and Control
Answer: C

16. Determine the angle of contact between the rope and the pulley. The rope is on a perfectly circular pulley, and either side of the rope extends at a certain angle from the horizontal from the pulley. From this, it can be discerned that the rope is in contact with the pulley for half of a rotation of a circle minus some amount based on the angle of entry and exit. Drawing a right triangle from the rope exiting at a 20° angle, it is determined that the difference between the angle of entry and angle of exit is 35°. This amount will be subtracted from the angle of half the rotation of a circle, which is 180°.

$$\theta = \left(180° - (55° - 20°)\right)\left(\frac{\pi \text{ radians}}{180°}\right) = 2.53 \text{ radians}$$

The angle of contact is now known, and the coefficient of friction was given in the problem statement. Use the relationship for belt friction to determine the output force.

$$F_1 = F_2 e^{\mu\theta} = (120 \text{ lb})e^{(0.22)(2.53 \text{ rad})} = 209 \text{ lb } (\sim 210 \text{ lb})$$

Reference: NCEES *FE Reference Handbook* > Statics > Belt Friction
Answer: B

17. The four resistors in the network are recognized as forming a Wheatstone bridge. From the perspective of the supply voltage (VX) resistors $R_1$and R_2 are in series, having an equivalent resistance of $R_{\text{eq1}} = R_1 + R_2 = 2\ \text{k}\Omega$, and resistors R_3 and R_G are in series, having an equivalent resistance of $R_{\text{eq1}} = R_3 + R_G = 700\ \Omega$. The two sets of series resistors are in parallel, so the equivalent resistance (R_{eq}) can be determined from:

$$\frac{1}{R_{\text{eq}}} = \frac{1}{\frac{1}{R_{\text{eq1}}} + \frac{1}{R_{\text{eq1}}}}$$

Substituting numerical values for R_{eq1} and R_{eq2} into the equation and solving for R_{eq} gives $R_{\text{eq}} = 518.5\ \Omega$. The current ($I$) drawn by a single strain gauge module can be found from Ohm's law to be:

$$I = \frac{V}{R_{\text{eq}}} = \frac{2.5\ \text{V}}{518.5\ \Omega} = 4.82\ \text{mA}$$

The number of strain gauge modules that draw 25 mA can be calculated as:

$$n = \frac{25\ \text{mA}}{4.82\ \text{mA}} = 5.2$$

A fractional number of devices is not physically possible, so the maximum number of strain gauge modules that can be powered without exceeding the rated current limit is five.

Reference: NCEES *FE Reference Handbook* > Electrical and Computer Engineering > Resistors in Series and Parallel
Answer: C

18. Benefits:
\$2,000 per year for the first five years

\$1,000 per year for the next five years

Another way to look at the benefits:
\$1,000 per year for ten years, plus \$1,000 benefit per year for the first five years

Maximum investment = present worth (PW) of benefits
$= A\ (P/A, i\%, n\ ) + A\ (P/A, i\%, n)$

$= \$1{,}000(P/A, 6\%, 10) + \$1{,}000(P/A, 6\%, 5)$

$= \$1{,}000(7.3601) + \$1{,}000(4.2124)$

$= \$11{,}573$

Reference: NCEES *FE Reference Handbook* > Engineering Economics > Interest Rate Tables
Answer: B

19. Both the sample standard deviation and arithmetic mean must be known to calculate the sample coefficient of variation.

Reference: NCEES *FE Reference Handbook* > Engineering Probability and Statistics > Dispersion, Mean, Median, and Mode Values
Answer: A, B

20. Since X is uniformly distributed on the range [3, 6], its probability density function is constant on the range [3, 6]. That is, $f(x) = c$.

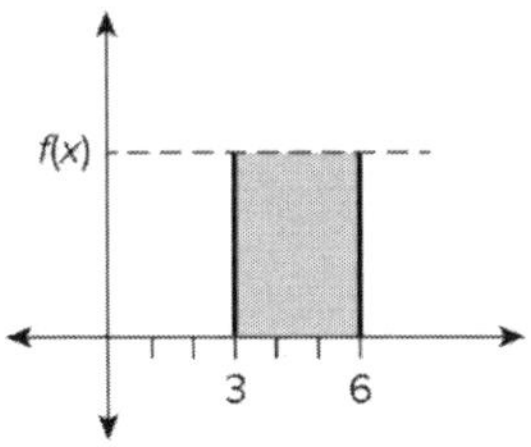

It is given that $P(3 \le X \le 6) = 1$.

$$P(3 \le X \le 6) = \int_3^6 f(x)dx = 1$$

$$\int_3^6 c dx = c(6 - 3) = 3c, \text{so } 3c = 1.$$

$c = 1/3$

$f(4) = c = 1/3$

Reference: NCEES *FE Reference Handbook* > Engineering Probability and Statistics > Probability Functions, Distributions, and Expected Values
Answer: A

21. Answers A and B: Saying nothing about the error is wrong. If the water line bursts, the engineer is to blame. This is also unethical.
Answer C: This is the correct answer. The best solution, although it costs more money, is to fix the problem.
Answer D: This answer is rash and is not the correct solution.

Reference: NCEES *FE Reference Handbook* > Ethics and Professional Practice > *Model Rules*, Section 240.15 Rules of Professional Conduct
Answer: C

22. Sum the moments about point A:

$$M_{R_A} = \sum M_A$$

$$M_{R_A} = 50 \text{ lb}(17 \text{ ft}) - 120 \text{ lb}(28 \text{ ft})$$

$$M_{R_A} = 2{,}510 \text{ ft} \cdot \text{lb}$$

Reference: NCEES *FE Reference Handbook* > Statics > Systems of Forces
Answer: D

23. $F = kx$

$$k = \frac{F}{x} = \frac{6 \text{ N}}{0.25 \text{ m}} = 24 \text{ N/m}$$

$$T = 2\pi\sqrt{\frac{m}{k}} = 2\pi\sqrt{\frac{3 \text{ kg}}{24 \text{ N/m}}} = 2.2 \text{ s}$$

Reference: NCEES *FE Reference Handbook* > Mechanical Engineering > Mechanical Design and Analysis
Answer: C

24. Calculate the tensile stress:

$$\sigma = \frac{P}{A} = \frac{200 \text{ kN}}{0.04 \text{ m}^2} = 5 \text{ MPa}$$

Calculate the strain:

$$\varepsilon = \frac{\delta}{L} = \frac{0.0004 \text{ m}}{8 \text{ m}} = 0.00005$$

Calculate the modulus of elasticity:

$$E = \frac{\sigma}{\varepsilon} = \frac{5 \text{ MPa}}{0.00005} = 100 \text{ GPa}$$

Determine the material: Cast iron (E = 100 GPa)

Reference: NCEES *FE Reference Handbook* > Mechanics of Materials > Uniaxial Loading and Deformation
Answer: C

25. A concurrent force system is one in which all of the lines of action of the applied forces meet at one point. By inspection, only the lines of action for figure B intersect at a common point.

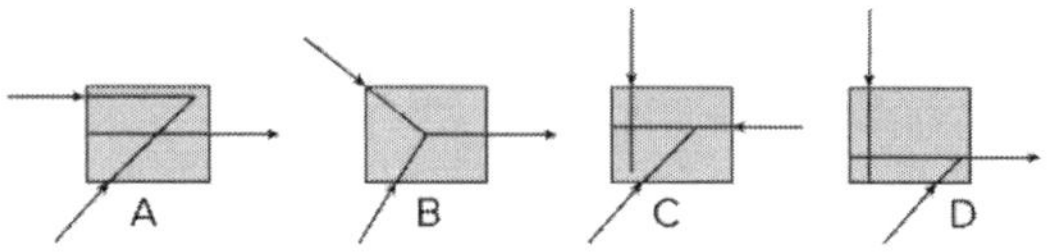

Reference: NCEES *FE Reference Handbook* > Statics > Concurrent Forces
Answer: B

26. Left-side loop, applying Kirchhoff's voltage rule (sum of voltage rises equals sum of voltage drops):
Counterclockwise loop (left side)
The sum of potential differences in the loop must equal zero (ΣV = 0).

$$-I_1(5\ \Omega) + 5\ \text{V} + 4\ \text{V} = 0$$

$$I_1 = \frac{-9\ \text{V}}{-5\ \Omega} = 1.8\text{A}$$

Right-side loop, applying Kirchhoff's voltage rule:
Clockwise loop (right side)

$$7\ \text{V} + 4\ \text{V} - I_2(3\ \Omega) = 0$$

$$I_2 = \frac{-11\ \text{V}}{-3\ \Omega} = 3.67\text{A}$$

Applying Kirchhoff's node/junction rule (sum of the currents entering a node equals sum of the currents exiting a node):

$$I_1 + I_2 + I_3 = 0$$

$$1.8\text{A} + 3.67\text{A} + I_3 = 0$$

$$I_3 = -5.47\text{A}$$

A negative sign indicates that the current is flowing in the opposite direction from what is shown in the figure.

Reference: NCEES *FE Reference Handbook* > Electrical and Computer Engineering > Kirchhoff's Laws
Answer: D

27. The normal stress in a beam due to bending can be calculated as:

$$\sigma_x = -\frac{My}{I}$$

On the neutral axis, $y = 0$.

$$\sigma_x = 0$$

Reference: NCEES *FE Reference Handbook* > Mechanics of Materials > Stresses in Beams
Answer: B

28. $a_x = \frac{F_x}{m}$

Where:
a_x = acceleration of the particle

F_x = the resultant of the applied forces

m = mass of the object

Note: g_c must be used to cancel out lbm/lbf and yield the proper unit of ft/s^2.

$$a_x = \frac{100 \text{ lbf}}{100 \text{ lbm}} \times \frac{1 \text{ lbm} \cdot \text{ft}}{32.2 \text{ lbf} \cdot \text{s}^2}$$

$$a_x = \frac{1 \text{ ft}}{32.2 \text{ s}^2} = 0.03 \text{ ft/s}^2$$

Reference: NCEES *FE Reference Handbook* > Dynamics > One-Dimensional Motion of a Particle (Constant Mass)
Answer: A

29. The probability density function of a *t*-distribution is symmetric about 0.

Therefore:
$$P(T > 0) = P(T < 0) = 1 - P(T > 0)$$

or

$$2P(T > 0) = 1$$

$$P(T > 0) = 0.50$$

Reference: NCEES *FE Reference Handbook* > Engineering Probability and Statistics > *t*-Distribution
Answer: B

30. As the client's representative, the engineer should not accept any gifts because the items could be construed as bribery. However, since the same bid was the lowest bid in a competitively bid project, that company must be awarded the contract.

Reference: NCEES *FE Reference Handbook* > Ethics and Professional Practice > *Model Rules,* Section 240.15 Rules of Professional Conduct
Answer: B

31. Trapezoidal rule (n = 1)

$$I = \int_a^b f(x)dx = \left(\frac{\Delta x}{2}\right)[f(a) + f(b)]$$

$$a = 0, b = 0.6, n = 1, f(x) = \int_0^{0.6}(\cos(x) + x)dx$$

With the trapezoidal rule:

$$\Delta x = (b - a) = 0.6 - 0 = 0.6$$

$$f(x) = (\cos(x) + x)$$

$$f(0) = 1; \; f(0.6) = 1.59$$

$$I = \left(\frac{\Delta x}{2}\right)[f(a) + f(b)] = 0.3\ (1 + 1.59) = 0.777$$

Reference: NCEES *FE Reference Handbook* > Mathematics > Numerical Integration
Answer: C

32. $C = \frac{\varepsilon A}{d} = \frac{\varepsilon_r\, \varepsilon_o A}{d}$

$$C_{air} = \frac{\varepsilon_{r,air}\, \varepsilon_o A}{d} \Rightarrow \frac{C_{air}}{\varepsilon_{r,air}} = \frac{\varepsilon_o A}{d}$$

$$C_4 = \frac{\varepsilon_{r,4}\, \varepsilon_o A}{d} \Rightarrow \frac{C_4}{\varepsilon_{r,4}} = \frac{\varepsilon_o A}{d}$$

$$\frac{C_{air}}{\varepsilon_{r,air}} = \frac{C_4}{\varepsilon_{r,4}}$$

$$C_4 = \frac{C_{air}}{\varepsilon_{r,air}} \varepsilon_{r,4} = \left(\frac{6}{1}\right)(4) = 24\ \mu\text{F}$$

Reference: NCEES *FE Reference Handbook* > Electrical and Computer Engineering > Capacitors and Inductors
Answer: D

33. First, determine the distance from the origin to the force vector. The distance to the point from the origin and the angle between this vector and the force vector are known. Using a right triangle, the distance from the origin to the force vector can be calculated. The distance will be the sin of the angle between the force and distance vectors. See the diagram below.

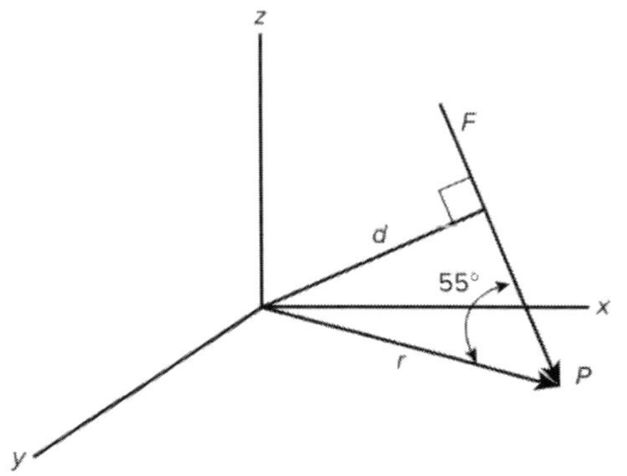

Calculate the distance (d) and then determine the magnitude of the moment about the origin. The distance will be the sin of the angle times the distance (r).

$d = r\sin(55)$
$d = 12\sin(55) = 9.83 \text{ ft}$

The moment about the origin is this distance times the magnitude of the force.
$M_0 = (9.83 \text{ ft})(60 \text{ kips}) = 589.8 \text{ kip} \cdot \text{ft} \approx 590 \text{ kip} \cdot \text{ft}$

Reference: NCEES *FE Reference Handbook* > Statics > Systems of Forces
Answer: B

34. In the equation below, $d = 1 \text{ m} + 0.5 \text{ m} = 1.5 \text{ m}$ and $h = d\sin(\theta)$.

$\frac{1}{2}kx^2 = mgh + mg\cos(\theta)\mu d$

Therefore:

$$\mu = \frac{\frac{1}{2}kx^2 - mgd\sin(\theta)}{mg\cos(\theta)d}$$

$$\mu = \frac{\frac{1}{2}400 \text{ N/m } (0.5 \text{ m})^2 - 5 \text{ kg } (9.8 \text{ m/s}^2)1.5 \text{ m } \sin(20°)}{5 \text{ kg } (9.8 \text{ m/s}^2)\cos(20°)\ 1.5 \text{ m}} = 0.36$$

Reference: NCEES *FE Reference Handbook* > Dynamics > Potential Energy
Answer: C

35. The material, loads, and cross sections are the same, so the strains in the two bars are the same. $\varepsilon = \frac{\Delta L}{L_o}$, so the elongation and the original length are proportional if the strains are the same. Therefore, the length of bar A is three times the length of bar B.

Reference: NCEES *FE Reference Handbook* > Mechanics of Materials > Engineering Strain
Answer: C

36. Plot the free-body diagram of joint D as shown below and assume the unknown member forces are tensile forces. Set the Cartesian coordinate system with the original point at joint D.

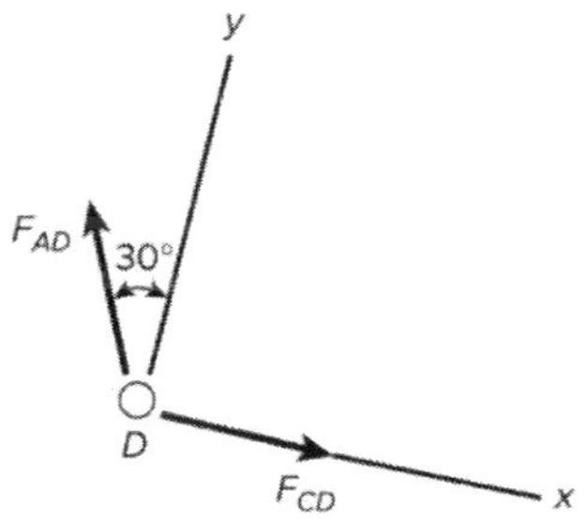

Apply the force equilibrium equations.

$\sum F_y = 0 \rightarrow F_{AD}\cos(30°) = 0 \rightarrow F_{AD} = 0$

$F_{CD} = 0$

There is no foundation for the other answer choices.

Reference: NCEES *FE Reference Handbook* > Statics > Systems of Forces
Answer: A

37. Find the shear strain (γ).

$$\gamma = \frac{\Delta}{\text{height}}$$

$$\gamma = \frac{0.03 \text{ mm}}{15 \text{ cm}} = 0.0002$$

Find the shear stress (τ).

$$\tau = \frac{T}{A}$$

$$\tau = \frac{20{,}000 \text{ N}}{0.2 \text{ m} \times 0.012 \text{ m}} = 8{,}333.3 \text{ kPa}$$

Find the shear modulus (G).

$$G = \frac{\tau}{\gamma}$$

$$G = \frac{8{,}333.3 \text{ kPa}}{0.0002} = 41{,}666{,}500 \text{ kPa} \times \frac{1 \text{ MPa}}{10^3 \text{ kPa}} = 41{,}666.50 \text{ MPa}$$

Reference: NCEES *FE Reference Handbook* > Mechanics of Materials > Shear Stress-Strain
Answer: C

38. $T = 2\pi\sqrt{\frac{m}{k}}$

k is calculated as:

$$0.2 = 2\pi\sqrt{\frac{0.5}{k}}$$

$$k = 493.5 \text{ N/m}$$

Reference: NCEES *FE Reference Handbook* > Dynamics > Free and Forced Vibration
Answer: D

39. Use the equation for AC motors:

$$n_s = 120\left(\frac{f}{p}\right)$$

$$f = \frac{n_s p}{120} = 1{,}800 \text{ rev/min} \times \frac{1 \text{ min}}{60 \text{ s}} \times \frac{2\pi}{\text{rev}} \times \frac{3}{120} = 4.712\left(\frac{1}{\text{s}}\right)$$

Equation for period:

$$T = \frac{1}{f} = \frac{1}{4.712\left(\frac{1}{\text{s}}\right)} = 0.212 \text{ s}$$

Reference: NCEES *FE Reference Handbook* > Electrical and Computer Engineering > AC Machines
Answer: A

40. The elements *i* and *j* represent the components of a force in each of the global axes for a system. The total magnitude of the force will be dependent on both of these components. Determine the magnitude using the Pythagorean theorem. The components of the force are each in line with the global axis, which forms two sides of a right triangle. The magnitude will be the longest side of the right triangle across from the right angle. This allows us to apply the Pythagorean theorem to determine the actual magnitude of the force.

$$F = \sqrt{(35 \text{ kips})^2 + (65 \text{ kips})^2} = 73.8 \text{ kips}$$

Reference: NCEES *FE Reference Handbook* > Statics > Resultant (Two Dimensions)
Answer: D

41. Find the Poisson's ratio.

Poisson's ratio: $v = -\frac{\varepsilon_{lateral}}{\varepsilon_{long}}$

$$\varepsilon_{lateral} = \frac{(D' - D)}{D} = \frac{0.020045\text{ m} - 0.02\text{ m}}{0.02\text{ m}} = 0.00225\text{ m}$$

$$\varepsilon_{long} = \frac{L' - L}{L} = \frac{5.946\text{ m} - 6\text{ m}}{6\text{ m}} = -0.009\text{ m}$$

$$v = \frac{-0.00225\text{ m}}{-0.009\text{ m}}$$

$$v = 0.25$$

Reference: NCEES *FE Reference Handbook* > Mechanics of Materials > Shear Stress-Strain
Answer: B

42. The relative velocity for impacts is:

$$e = \frac{v_2' - v_1'}{v_1 - v_2}$$

Where:
v_1 = initial velocity of the particle

v_1' = final velocity of the particle

v_2 = initial velocity of the rigid object

v_2' = final velocity of the rigid object

Since the collision has been defined as perfectly elastic ($e = 1$) and the rigid object has been defined as not movable, v_2 and v_2' are both equal to zero and the above equation simplifies to:

$$e = \frac{0 - v_1'}{v_1 - 0} \rightarrow v_1 e = -v_1'$$

Since kinetic energy is equal to $\frac{mv^2}{2}$, the particle will retain all the kinetic energy and will rebound.

Reference: NCEES *FE Reference Handbook* > Dynamics > Impulse and Momentum
Answer: B

43. $d = \sqrt{(X_B - X_A)^2 + (Y_B - Y_A)^2 + (Z_B - xZ_A)^2}$

$d = \sqrt{(3-4)^2 + (8-7)^2 + (5-8)^2}$

$d = \sqrt{1+1+9} = 3.3$

Reference: NCEES *FE Reference Handbook* > Mathematics > Quadric Surface (SPHERE)
Answer: B

44. Convert the $300 monthly payments to equivalent annual payments.

Monthly interest = 4%/12 = 0.33% or 0.0033

Equivalent annual payment = $300(*F*/*A*, *i*%, *n*)

$= (\$300)\dfrac{(1+i)^n - 1}{i}$

$= \left(\$300\dfrac{(1+0.0033)^{12} - 1}{0.0033}\right)$

$= \$3{,}666$

With the above operations, 120 monthly payments of $300 were converted to ten equivalent annual payments of $3,666.

Find the present value of new payments:

$P = \$3{,}666(P/A, 4\%, 10)$

= $3,666(8.1109)

= $29,735

Reference: NCEES *FE Reference Handbook* > Engineering Economics > Factor Name table, Interest Rate Tables
Answer: C

45. The equation below defines the radius of gyration about the *x* axis.

$$r_x = \sqrt{\frac{I_x}{A}}$$

Use the parallel-axis theorem to determine the axial moment of inertia about the *x* axis.

Substitute for the centroidal axial moment of inertia, area, and *x* distance between the axes. The area will be 7 cm × 4 cm, and the distance from the *x* axis *dy* will be 9 cm.

$$I_x = I_{x_c} + A\,dy^2$$

$$I_x = \frac{1}{12}(7\text{ cm})(4\text{ cm})^3 + (7\text{ cm})(4\text{ cm})(9\text{ cm})^2$$

$$= 2{,}305.33\text{ cm}^4$$

Substitute the moment of inertia and area into the equation for the radius of gyration.

$$r_x = \sqrt{\frac{2{,}305.33\text{ cm}^4}{(7\text{ cm})(4\text{ cm})}}$$

$$r_x = 9.07\text{ cm}$$

Reference: NCEES *FE Reference Handbook* > Statics > Radius of Gyration
Answer: C

46. Answer A is correct: The deflection at the free end is maximum.
Answer B is not correct: The slope at the free end is maximum, and it is negative because the angle is clockwise.
Answer C is correct: The slope at the free end is maximum, and it is negative.
Answer D is correct: Both answers A and C are correct.

Reference: NCEES *FE Reference Handbook* > Mechanics of Materials > Cantilevered Beam Slopes and Deflections
Answer: B

47. $\omega_n = \sqrt{\frac{k}{m}}$

$K = \frac{W}{\delta_{st}}$

In the above equation, W is the weight of the object that is equal to the spring force at the static equilibrium.

$M = \frac{W}{g}, \quad \omega_n = \sqrt{\frac{g}{\delta_{st}}}$

Reference: NCEES *FE Reference Handbook* > Dynamics > Free and Forced Vibration
Answer: A

48. (1) $\sigma = \frac{F}{A}$

Where:
σ = engineering stress

F = applied force

A = initial cross-sectional area

(2) $\delta = \frac{PL}{AE}$

Where:
$\delta = \Delta L$ = change in length

L = length

P = force

A = area

E = modulus of elasticity

Combine to give:

$\delta = \frac{\sigma L}{E} = \frac{(350 \times 10^6\ \text{N/m}^2)(3\ \text{m})}{70 \times 10^9\ \text{N/m}^2} = 15\ \text{mm}$

Reference: NCEES *FE Reference Handbook* > Materials Science/Structure of Matter > Engineering stress, Hooke's Law
Answer: D

49. $\sum F_x = 0 \Rightarrow -F_{CD}\sin 45° + F = 0 \Rightarrow F_{CD} = 707\text{ lb}$

$\sum F_y = 0 \Rightarrow F_{CD}\cos 45° + F_{BD} = 0 \Rightarrow F_{BD} = -500\text{ lb}$

The force in member BD is 500 lb (compression).

Reference: NCEES *FE Reference Handbook* > Statics > Systems of Forces
Answer: A

50. A restoring force is always one that is opposite in direction to displacement and also obeys Hooke's law.

Answer: A, D

51. Develop a free-body diagram for this system. The force on one side of a pulley must be conserved on the other side. Any support for the pulley will take the force of both sides of the pulley. The force (T) on the top right pulley will have an equivalent force (T) on the rope on the other side.

The fixed ceiling will support a load for either side of the pulley, which will be d ouble the applied tension force. The next pulleys attached to the one at the top right will also carry this tension force. Draw a free-body diagram to represent this system.

Having traced the tension load to the block, the force required (T) can be dete rmined knowing the block weight. Sum the forces vertically to determine the force (T).

$$\sum F = 0 = \frac{1}{2}T + \frac{1}{2}T - 150\text{ kips}$$

$T = 150$ kips

Reference: NCEES *FE Reference Handbook* > Statics > Systems of Forces
Answer: D

52. Use the common projectile motion formula obtained from the constant acceleration equation shown below. It is stated that Eric is already standing 250 ft behind the cannons, so only the time needed for the projectiles to travel 2,250 ft is needed.

$$x = v_o \cos(\theta) t + x_o$$

$$x = 2{,}250 \text{ ft}$$

$$2{,}250 \text{ ft} = (2{,}500 \text{ ft/s}) \cos(40°)\, t + 0$$

Then, solve for t:

$$t = \frac{2{,}250 \text{ ft}}{(2{,}500 \text{ ft/s})\cos(40°)} = 1.17 \text{ s}$$

Reference: NCEES *FE Reference Handbook* > Dynamics > Projectile Motion
Answer: B

53. The eccentricity of a hyperbola is always greater than 1.

Reference: NCEES *FE Reference Handbook* > Mathematics > Case 3. Hyperbola
$e > 1$
Answer: Hyperbola

54. Calculate the rotational speed of the flywheel:

$$\omega_o = \frac{650 \text{ rev/min } (2\pi \text{ rad/rev})}{60 \text{ s/min}}$$

$$\omega_o = 68.03 \text{ rad/s}$$

The overall displacement can then be calculated with the uniform acceleration formula:

$$\theta = 0.5t(\omega + \omega_o)$$

$$\theta = 0.5(0.5 \text{ s})(0 + 68.03 \text{ rad/s})$$

$$\theta = 17.0075 \text{ rad}$$

Convert radians to revolutions, where 1 rad = 0.159 rev.
0.159 (17.0075) = 2.7 rev

Reference: NCEES *FE Reference Handbook* > Dynamics > Constant Acceleration
Answer: B

55. Plastic deformation and fibrous fracture surfaces are signs of ductile failure. Elastic deformation is not a sign of failure. Therefore, a cleaved fracture surface is a sign of brittle failure.

Answer: B

56. Engineering stress, $\sigma = F/A$ (initial) = 62,255 lbf/0.52 in^2 (≈ 119,721 psi)

Using Hooke's Law, $\sigma = E\,\varepsilon$

$E = \sigma/\varepsilon$ = 119,721 psi/0.0041 in/in = 29,200,244 psi

Reference: NCEES *FE Reference Handbook* > Materials Science/Structure of Matter > Properties of Materials – Mechanical
Answer: C

57. $P_{\text{atm}} = P_V + \gamma h$

$$h = \frac{P_{\text{atm}} - P_V}{\gamma} = \frac{101.3 \text{ kPa} - 2.3 \text{ kPa}}{8{,}703 \text{ N/m}^3} = 11.4 \text{ m}$$

Reference: NCEES *FE Reference Handbook* > Fluid Mechanics > Manometers
Answer: B

58. According to the Gibbs phase rule, $P + F = C + 2$.

Where:
P = total number of phases

F = degrees of freedom

C = number of components

$P = 3$ and $C = 2$

$F = C + 2 - P$

$F = 2 + 2 - 3$

$F = 1$

Reference: NCEES *FE Reference Handbook* > Thermodynamics > Phase Relations
Answer: 1

59. Fourier's law of conduction states that the rate of heat transfer through an area is proportional to the temperature difference along a line normal to that area.

$$Q = -kA\left(\frac{dT}{dx}\right)$$

Where:

Q = rate of heat transfer

A = area

k = thermo conductivity of medium (proportionality constant)

$\frac{dT}{dx}$ = differential change in temperature

Reference: NCEES *FE Reference Handbook* > Heat Transfer > Basic Heat-Transfer Rate Equations
Answer: D

60. Using the damped natural frequency formula:

$$\omega_d = \omega_n \sqrt{1-\zeta^2}$$

$$\omega_d = (32)\sqrt{1-(0.76)^2}$$

$$\omega_d = 20.8$$

Reference: NCEES *FE Reference Handbook* > Instrumentation, Measurement, and Control > Control Systems
Answer: D

61. According to the Coulomb-Mohr theory, when a member subject to a general state with the principal stresses of $\sigma_1 \geq \sigma_2 \geq \sigma_3$ in which $\sigma_1 > 0$ and $\sigma_3 < 0$, the material failure occurs whenever:

$$\frac{\sigma_1}{S_{ut}} - \frac{\sigma_3}{S_{uc}} \geq 1$$

In this problem, the member has $S_{ut} = 50$ kpsi and $S_{uc} = 150$ kpsi. It is subject to the general stress state with the principal stresses $\sigma_1 = 30$ kpsi, $\sigma_2 = -20$ kpsi, and $\sigma_3 = -40$ kpsi.

$$\frac{\sigma_1}{S_{ut}} - \frac{\sigma_3}{S_{uc}} = \frac{30}{50} - \left(-\frac{40}{150}\right) = 0.87 < 1$$

Reference: NCEES *FE Reference Handbook* > Mechanical Engineering > Static Loading Failure Theories
Answer: C

62. In general, materials with higher fracture toughness exhibit better fatigue properties. 4340 steel has the greatest fracture toughness value.

Reference: NCEES *FE Reference Handbook* > Materials Science/Structure of Matter > Representative Values of Fracture Toughness
Answer: B

63. According to the P-V diagram of a Carnot cycle, the four processes are explained below.

- During process 1-2, gas is compressed adiabatically from state 1 to state 2 where the temperature reaches T_H.
- During process 2-3, gas expands isothermally at T_H until it reaches state 3.
- During process 3-4, gas expands adiabatically until state 4 where the temperature drops to T_L.
- During process 4-1, gas is compressed isothermally at T_L until it reaches state 1.
- So, at process 3-4, the system is experiencing an adiabatic expansion process.

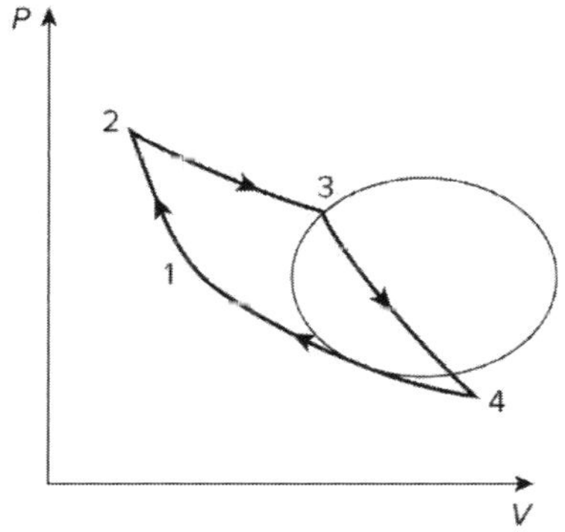

Reference: NCEES *FE Reference Handbook* > Thermodynamics > Common Thermodynamic Cycles

Answer: Circle process 3-4

64. Using the analog-to-digital converter resolution equation:

$$\varepsilon_v = \frac{V_H - V_L}{2^n} = \frac{120\text{ V} - 12\text{ V}}{2^4} = 6.75$$

Reference: NCEES *FE Reference Handbook* > Instrumentation, Measurement, and Control > Analog-to-Digital Conversion

Answer: C

65. Biot number:

$$\mathrm{Bi} = \frac{hV}{kA_S}$$

$$\frac{V}{A_S} = \frac{\pi r^2 l}{2\pi r l + \pi r^2} = \frac{rl}{2l + r} = \frac{0.002(0.01)}{2(0.01) + 0.002} = 9.1 \times 10^{-4}\ \mathrm{m}$$

$$\mathrm{Bi} = \frac{85(9.1 \times 10^{-4})}{8.1} = 9.5 \times 10^{-3}$$

Reference: NCEES *FE Reference Handbook* > Heat Transfer > Transient Conduction Using the Lumped Capacitance Model
Answer: D

66. $h = \frac{4\sigma\cos(\beta)}{\gamma d} = \frac{4(0.0728\ \mathrm{N/m})\cos(0^\circ)}{(9{,}789\ \mathrm{N/m^3})d} = 0.003\ \mathrm{m}$

$$d = \frac{4\sigma\cos(\beta)}{\gamma h} = \frac{4(0.0728\ \mathrm{N/m})\cos(0^\circ)}{(9{,}789\ \mathrm{N/m^3})(0.003\ \mathrm{m})} = \frac{0.2912}{29.367}$$

$$= 0.00992\ \mathrm{m} = 9.92\ \mathrm{mm}$$

Reference: NCEES *FE Reference Handbook* > Fluid Mechanics > Surface Tension and Capillarity
Answer: B

67. According to the *FE Reference Handbook*, a location interference fit (H7/p6) should be used for "parts requiring rigidity and alignment with prime accuracy of location but without special bore pressure requirements."

Reference: NCEES *FE Reference Handbook* > Mechanical Engineering > Some Preferred Fits
Answer: D

68. Module = $m = d/N$

Generally, the design is based on the weaker element, which is commonly the pinion:

$d = mN = 4\text{ mm}(20) = 80\text{ mm}$

Circular pitch = $pc = \frac{\pi D}{N} = \frac{\pi 80}{20} = 12.56\text{ mm}$

Base pitch = $pb = pb\cos(\alpha)$

Where:
α = pressure angle (20°)

$pb = 12.56\cos(20) = 11.8\text{ mm}$

The length between shafts is called center distance.

$$C = \frac{d_1 + d_2}{2}$$

Where:

d_1 = pitch circle diameter of pinion (80 mm)

d_2 = pitch circle diameter of gear = $mN\left(m = \frac{d_2}{N}\right)$

$d_2 = mN = 4(80) = 320\text{ mm}$

$$C = \frac{80\text{ mm} + 320\text{ mm}}{2} = 200\text{ mm}$$

Reference: NCEES *FE Reference Handbook* > Mechanical Engineering > Power Transmission
Answer: C

69. $F_D = \frac{C_D \rho v^2 A}{2}$

At terminal velocity, $F_D = \text{weight} = mg$

$$F_D = \frac{C_D \rho v^2 A}{2} = mg$$

$$mg = (v\rho_{\text{sphere}})g = \left(\frac{4}{3}\pi\frac{r^3}{3}\rho_{\text{sphere}}\right)$$

$$A = \pi(d/2)^2$$

$$\left(\frac{1}{2}\right) C_D \rho_{\text{air}} v_t^{\,2} \pi \left(\frac{d}{2}\right)^2 = \frac{4}{3}\pi\left(\frac{d}{2}\right)^3 \rho_{\text{sphere}} g$$

$$v_t = \sqrt{\frac{(2)(4\pi)\left(\frac{d}{2}\right)^3 \rho_{\text{sphere}} g}{3 C_D \rho_{\text{air}} \pi \left(\frac{d}{2}\right)^2}} = \sqrt{\frac{4 d \rho_{\text{sphere}} g}{3 C_D \rho_{\text{air}}}}$$

$$v_t = \sqrt{\frac{(4)(0.044 \text{ m})(2{,}340 \text{ kg/m}^3)(9.81 \text{ m/s}^2)}{(3)(0.6)(1.225 \text{ kg/m}^3)}}$$

v_t = 42.8 m/s

Reference: NCEES *FE Reference Handbook* > Fluid Mechanics > Drag Force
Answer: C

70. Use this formula to calculate the largest possible crack size:

$$K_{IC} = Y\sigma\sqrt{\pi a}$$

Fracture toughness of 4340 steel is 42 ksi-in$^{0.5}$. Reordering the formula above, solve for the crack size (a).

$$a = \frac{\left(\frac{K_{IC}}{Y\sigma}\right)^2}{\pi} = \frac{\left(\frac{42 \text{ ksi-in}^{0.5}}{1.1(50 \text{ ksi})}\right)^2}{\pi} = 0.186 \text{ in}$$

Reference: NCEES *FE Reference Handbook* > Materials Science/Structure of Matter > Fracture Toughness
Answer: A

71. Energy $=$ (Power)(Time)

$= (8\text{ kJ/s})(15\text{ min})(60\text{ s}/1\text{ min})$

$= 7{,}200\text{ kJ}$

Answer: D

72. Find the mean diameter of the spring coil (D) and the diameter of the spring wire (d). The spring is operated in a casing with a diameter of 36 mm. The outer diameter of the spring coil (D_2) is less than 36 mm.

$D_2 = D + d < 36\text{ mm}$

The deflection of the spring:

$F = kx \rightarrow x = F/k$

$$k = \frac{d^4 G}{8D^3 N}$$

$$x = \frac{8FC^3 N}{Gd}$$

$$C = \frac{D}{d} \rightarrow \frac{D^3}{d^4} = \frac{C^3}{d}$$

$$72 = \frac{(8 \times 870 \times C^3 \times 43)}{(87 \times 10^3 \times D)} = 3.44 \times \left(\frac{C^3}{D}\right)$$

$$\rightarrow \frac{C^3}{d} = 20.93$$

Neither C nor d is specified in the problem. Assume $d = 3$ mm.

$C^3 = 20.93(3) = 62.79 \rightarrow C = 3.97$

$D = Cd = 3.97(3) = 11.91\text{ mm}$

The outer diameter of the spring:

$D_2 = D + d = 11.91\text{ mm} + 3\text{ mm} = 14.91\text{ mm} < 36\text{ mm}$

The assumed $d = 3$ mm is correct.

Reference: NCEES *FE Reference Handbook* > Mechanical Engineering > Mechanical Design and Analysis

Answer: A

73. $\%\text{OS} = 100e^{-\pi\zeta/\sqrt{1-\zeta^2}}$

Solving with the given variables, $\%\text{OS} = 100e^{-\pi(0.84)/\sqrt{1-(0.84)^2}} = 0.77\%$

Reference: NCEES *FE Reference Handbook* > Instrumentation, Measurement, and Control > Second-Order Control System Models
Answer: B

74. Options A and B must be incorrect because thermal conductivity and heat capacity are material properties.

The thermal energy of the steel is the amount of heat the steel contains, or its temperature. Since the block is cooling, its thermal energy is decreasing, which means option C is incorrect.

Answer: D

75. Find the intersection of P = 0.1 MPa and x = 0.2. The intersection of the two lines is the status point. Interpolate between constant entropy lines of 1.05 and 1.1.

s = 1.08 kJ/kg·K

Reference: NCEES *FE Reference Handbook* > Thermodynamics > P-*h* Diagram for Refrigerant HFC-134a
Answer: B

76. $\tau = K\left(\frac{dv}{dy}\right)^n$

$$\frac{31\ \text{N}}{(0.6)(0.35)} = 0.44\left(\frac{8}{0.002}\right)^n = 0.44(4{,}000)^n$$

$n = 0.70$

Reference: NCEES *FE Reference Handbook* > Fluid Mechanics > Stress, Pressure, and Viscosity
Answer: B

77. $\varepsilon_{th} = \alpha\,(T_2 - T_1)$

$\varepsilon_{th} = 12.8 \times 10^{-6}\,(800° - 70°) = 0.00934$

$\varepsilon_{th} = 9.34 \times 10^{-3}$

Reference: NCEES *FE Reference Handbook* > Materials Science/Structure of Matter > Thermal Properties
Answer: A

78. $f_s = \frac{1}{\Delta t}$

The time period is given as 0.0083 s, so the sampling rate is:

$$f_s = \frac{1}{0.0083\ \text{s}} = 120\ \text{Hz}$$

Reference: NCEES *FE Reference Handbook* > Instrumentation, Measurement, and Control > Sampling
Answer: B

79. According to the distortion-energy theory, when a member is subject to a general state with the principal stresses of $\sigma_1 \geq \sigma_2 \geq \sigma_3$, the material failure occurs when $\sigma' \geq S_y$.

The von Mises stress is $\sigma' = \left[\frac{(\sigma_1-\sigma_2)^2+(\sigma_2-\sigma_3)^2+(\sigma_1-\sigma_3)^2)}{2}\right]^{\frac{1}{2}}$.

In this problem, the member has $S_y = 60$ kpsi. It is subject to the general stress state with the principal stresses $\sigma_1 = 30$ kpsi, $\sigma_2 = -20$ kpsi, and $\sigma_3 = -40$ kpsi.

$$\sigma' = \left[\frac{(50)^2 + (20)^2 + (70)^2}{2}\right]^{\frac{1}{2}}$$

$\sigma' = 62.4\ \text{kpsi} > 60\ \text{kpsi}$

Reference: NCEES *FE Reference Handbook* > Mechanical Engineering > Distortion-Energy Theory
Answer: A

80. $\text{Re} = \frac{vD\rho}{\mu} = \frac{vD}{\nu}$

$$D = 4\left(\frac{\text{Cross-sectional area}}{\text{Wetted perimeter}}\right) = 4\left(\frac{(1.7\text{ m})(3\text{ m})}{1.7\text{ m} + 3\text{ m} + 1.7\text{ m}}\right) = 3.2\text{ m}$$

$$Q = vA$$

$$v = \frac{Q}{A} = \frac{(1\text{ m}^3/\text{s})}{(1.7\text{ m} \times 3\text{ m})} = 0.196\text{ m/s}$$

$$\text{Re} = \frac{vD}{\nu} = \frac{(0.196\text{ m/s})(3.2\text{ m})}{0.000000800\text{ m}^2/\text{s}} = 784{,}000 = 7.84 \times 10^5$$

Reference: NCEES *FE Reference Handbook* > Fluid Mechanics > Reynolds number, Properties of Water
Answer: C

81. $\eta = 1 - r^{1-k}$, where $r = \frac{v_1}{v_2}$, $k = \frac{c_p}{c_v} = 1.40$

$$r = \frac{v_1}{v_2} = \frac{p_2}{p_1}$$

$$\eta = 1 - \left(\frac{p_2}{p_1}\right)^{1-k}$$

$$-\eta + 1 = \left(\frac{p_2}{p_1}\right)^{1-k}$$

$$(-\eta + 1)^{\frac{1}{1-k}} = \frac{p_2}{p_1}$$

$$p_1 = \frac{p_2}{(-\eta + 1)^{\frac{1}{1-k}}}$$

$$p_1 = \frac{16\text{ bars}}{(-0.5 + 1)^{\frac{1}{1-1.40}}}$$

$$p_1 = 2.83\text{ bars} = 283\text{ kPa}$$

Reference: NCEES *FE Reference Handbook* > Thermodynamics > Common Thermodynamic Cycles
Answer: B

82. A strain gauge is the only device listed here that can measure elongation of a material. The suitcase can be hung from the calibrated strain gauge to determine its weight.

Reference: NCEES *FE Reference Handbook* > Instrumentation, Measurement, and Control > Strain Transducers
Answer: A

83. The maximum distance is 5.125 in between the radii.

The smallest radius is 0.875 in.

5.125 – 0.875(2) = 3.375 in

Answer: B

84. $k = 1.07$ W/mK

$L = 0.25$ m

$A = 1.2$ m^2

Thermal resistance due to firebrick is calculated as:

$$R = \frac{L}{kA} = \frac{0.25}{1.07 \times 1.2} = 0.195 \text{ K/W}$$

Reference: NCEES *FE Reference Handbook* > Heat Transfer > Thermal Resistance (R)
Answer: A

85. During the cold-working process, the grains dislocate and move, generating additional dislocations that pile up. The more dislocations within the steel, the more they will interact and tangle, resulting in decreased mobility of the grains and increased strength.

Reference: NCEES *FE Reference Handbook* > Materials Science/Structure of Matter > Thermal and Mechanical Processing
Answer: A

86. $Q_1 = Q_2 + Q_3$

$800 \text{ gpm} = 500 \text{ gpm} + Q_3$

$Q_3 = 300 \text{ gpm}$

Reference: NCEES *FE Reference Handbook* > Thermodynamics > Steady-Flow Systems
Answer: C

87. Specific enthalpy, $\Delta h = c_p \Delta T$

= (2.05 kJ/kg·K) (215 K – 273 K)

= -119 kJ

Note that the final temperature is lower than the beginning temperature. Therefore, the change in enthalpy is negative.

Reference: NCEES *FE Reference Handbook* > Thermodynamics > State Functions (properties)
Answer: B

88. $Q = \frac{2\pi k L (T_1 - T_2)}{\ln\left(\frac{r_2}{r_1}\right)}$

$$Q = \frac{2(3.14)(285 \text{ W/m-k})(2 \text{ m})(45° - 15°)}{\ln\left(\frac{0.025}{0.02}\right)}$$

$Q = 481.3 \text{ kW}$

Reference: NCEES *FE Reference Handbook* > Heat Transfer > Conduction Through a Cylindrical Wall
Answer: B

89. According to the maximum-normal-stress theory, material failure occurs when the greatest positive principal stress exceeds the tensile strength or when the greatest negative principal stress exceeds the compressive strength. For plane stress with the principal stresses $\sigma_1 \geq \sigma_2 (\sigma_3 = 0)$, the maximum-normal-stress theory can be expressed as $\sigma_1 \geq S_{ut}$ or $\sigma_2 \leq -S_{uc}$.

If the member is subject to a plane stress state with the principal stresses of $0 > \sigma_1 \geq \sigma_2$, material failure will not occur if $\sigma_2 > -S_{uc}$.

In this problem, the member has $S_{uc} = 100$ kpsi, and it is subject to the following greatest negative principal stresses:

A. -180 kpsi
B. -130 kpsi
C. -90 kpsi
The answer is C since $\sigma_2(-90 \text{ kpsi}) > -S_{uc}(-100 \text{ MPa})$.

Reference: NCEES *FE Reference Handbook* > Mechanical Engineering > Static Loading Failure Theories
Answer: C

90. A high ductile-to-brittle transition temperature increases the minimum temperature at which a pressure vessel can be safely operated. This is not desirable. All other listed properties are desirable to have in pressure vessel construction.

Reference: NCEES *FE Reference Handbook* > Materials Science/Structure of Matter > Impact Test
Answer: D

91. Heat transfer is calculated as:

$$Q = \frac{2\pi L(T_i - T_o)}{\frac{\ln\left(\frac{r_2}{r_1}\right)}{k_A} + \frac{\ln\left(\frac{r_3}{r_2}\right)}{k_B}}$$

$$\frac{Q}{L} = \frac{2\pi(T_i - T_o)}{\frac{\ln\left(\frac{r_2}{r_1}\right)}{k_A} + \frac{\ln\left(\frac{r_3}{r_2}\right)}{k_B}}$$

$$\frac{Q}{L} = \frac{2 \times 3.14(100°\text{C} - 40°\text{C})}{\frac{\ln\left(\frac{0.05}{0.035}\right)}{15\ \text{W/m°C}} + \frac{\ln\left(\frac{0.07}{0.05}\right)}{0.08\ \text{W/m°C}}}$$

$$\frac{Q}{L} = 89.08\ \text{W/m}$$

Reference: NCEES *FE Reference Handbook* > Heat Transfer > Conduction
Answer: C

92. The bolt load on the bolt is calculated as:

$$F_b = CP + F_i$$

$$F_b = (0.168 \times 38\ \text{lbf}) + 70\ \text{lbf}$$

$$F_b = 76.384\ \text{lbf}$$

Reference: NCEES *FE Reference Handbook* > Mechanical Engineering > Joining Methods
Answer: B

93. Assume carbon monoxide and air is ideal gas. Then: $V_{CO} = \frac{n_{CO}RT}{P}, V_t = \frac{n_tRT}{P}$

The mole fraction of carbon monoxide: $y_{CO} = \frac{n_{CO}}{n_t} = \frac{V_{CO}}{V_t} = 0.04$

Partial pressure of monoxide in the mixture: $P_{CO} = Py_{CO} = 1(0.04) = 0.04$ atm

Reference: NCEES *FE Reference Handbook* > Thermodynamics > *PVT* Behavior
Answer: A

94. $v = k_1 \, C R_H^{0.63} S^{0.54}$

$$v = (1.318)(120)(0.75)^{0.63}\left(\frac{10.2 \text{ ft}}{2{,}000 \text{ ft}}\right)^{0.54} = 7.63 \text{ ft/s}$$

$$Q = Av = (\pi/4)(2)^2(7.63) = 24.0 \text{ ft}^3/\text{s}$$

Reference: NCEES *FE Reference Handbook* > Fluid Mechanics > Open-Channel Flow and/or Pipe Flow of Water
Answer: D

95. Fatigue failure is caused by repeated loading.

Reference: NCEES *FE Reference Handbook* > Materials Science/Structure of Matter > Mechanical
Answer: A

96. For a parallel flow concentric tube heat exchanger:

$$\varepsilon = \frac{1 - \exp[-NTU(1 + C_r)]}{1 + C_r}$$

$$\varepsilon = \frac{1 - \exp[-1.1 \times (1 + 0.8)]}{1 + 0.8} = 0.48$$

Reference: NCEES *FE Reference Handbook* > Heat Transfer > Effectiveness-NTU Relations
Answer: C

97. Find the joint coefficient C:

$$C = \frac{k_b}{k_b + k_m}$$

$$C = \frac{1.79}{(1.79) + (11.33)}$$

$$C = 0.136$$

The total bolt load is calculated as:

$$F_b = CP + F_i$$

$$F_i = F_b - CP$$

$$F_i = 80 - (0.136 \times 50)$$

$$F_i = 73.2 \text{ lbf}$$

The total load on the members is calculated as:

$$F_m = (1 - C)\, P - F_i$$

$$F_m = (1 - 0.136)(50) - 73.2 = -30 \text{ lbf}$$

Reference: NCEES *FE Reference Handbook* > Mechanical Engineering > Joining Methods
Answer: A

98. The Carnot cycle efficiency is originally:

$$\eta = \frac{T_H - T_L}{T_H}$$

$$\eta = \left(\frac{(2{,}500°\text{C} + 273) - (50°\text{C} + 273)}{2{,}773}\right) = 88\%$$

After the change:

$$\eta = \frac{(2{,}750°\text{C} + 273) - (130°\text{C} + 273)}{(2{,}750°\text{C} + 273)} = 86.7\% \text{ (efficiency decreases)}$$

Reference: NCEES *FE Reference Handbook* > Thermodynamics > Basic Cycles
Answer: B

99. $P_1 - P_2 = \gamma(z_1 - z_2) = \gamma h = \rho g h$

$P_1 + \gamma_w(z_1 - z_2) + SG_{Hg}\gamma_w z_3 - SG_{\text{gasoline}}\gamma_w(z_3 + z_a) = P_a$

$$101.3 \text{ kPa} + (9.81 \text{ kN/m}^3)(0.5 \text{ m} - 0.1 \text{ m}) - (13.6)(9.81 \text{ kN/m}^3)(0.09 \text{ m}) + (0.91)(9.81 \text{ kN/m}^3)(0.09 \text{ m} + 0.05 \text{ m}) = P_a$$

$P_a = 94.47$ kPa (abs)

Reference: NCEES *FE Reference Handbook* > Fluid Mechanics > Characteristics of a Static Liquid
Answer: B

100. $S_c = (S_f \times V_f) + (S_r \times V_r)$

$S_c = (375{,}000 \text{ psi} \times 0.72) + (8{,}000 \text{ psi} \times 0.28)$

$S_c = 272{,}240$ psi

Reference: NCEES *FE Reference Handbook* > Materials Science/Structure of Matter > Composite Materials
Answer: C

101. The force required to cause a spring to reach its yield strength is F in the max shear stress equation since design shear stress is nothing but shear stress:

$$\tau = K\left(\frac{8FC}{\pi d^2}\right)\left(\text{as } C = \frac{D}{d}\right)$$

$$K = \frac{2C + 1}{2C} \rightarrow \frac{[2(8) + 1]}{2(8)} = 1.0625$$

$$720 \text{ MPa} = 1.0625\left(\frac{8F(8)}{\pi(5)^2}\right) = 0.8658F$$

$$\frac{720 \text{ MPa}}{0.8658} = F$$

$F = 831.60$ N

Reference: NCEES *FE Reference Handbook* > Mechanical Engineering> Mechanical Design and Analysis
Answer: C

102. Based on energy balance, the conductive heat transfer is equal to the convective heat transfer at either surface. This is equal to the heat flux across:

$$Q_{\text{cond}} = Q_{\text{conv},1} = Q_{\text{conv},2} = Q$$

Determine the total thermal resistance. The plane wall conduction resistance is calculated as:

$$\frac{K}{W}: R = \frac{L}{kA}$$

Convective resistance is calculated as:

$$\frac{K}{W}: R = \frac{1}{hA}$$

$$R_1 = \frac{0.09}{13A} = \frac{0.0069}{A}\ \text{m}^2\text{K/W}$$

$$R_2 = \frac{1}{22A} = \frac{0.045}{A}\text{m}^2\text{K/W}$$

$$R_3 = \frac{1}{22A} = \frac{0.045}{A}\text{m}^2\text{K/W}$$

$$R_{\text{tot}} = R_1 + R_2 + R_3 = \frac{0.097}{A}\text{m}^2\text{K/W}$$

Determine the heat flux per area across the wall, then solve temperatures through convective heat transfer rates:

$$Q = \frac{\Delta T}{R_{\text{tot}}}$$

In the above equation, $\Delta T = (39°\text{C} - 11°\text{C})$

$$Q = \frac{28}{\frac{0.097}{A}} = 289A\ \text{W}$$

$$Q = Q_{\text{conv},1} = hA(T_{\infty,1} - T_1)$$

In the above equation, $T_{\infty,1} = 39°\text{C}, h = 22\ \text{W}/(\text{m}^2\text{K}), \text{and } T_1 = 26°\text{C}.$

Therefore:

$$Q = Q_{\text{conv},2} = hA(T_2 - T_{\infty,2})$$

In the above equation, $T_{\infty,2} = 11°\text{C};\ T_2 = 24°\text{C}$

Reference: NCEES *FE Reference Handbook* > Heat Transfer > Thermal Resistance (R)

Answer: B

103. For a double-pipe heat exchanger:

$$\dot{m}_F(h_{1i} - h_{1e}) = \dot{m}_W(h_{2e} - h_{2i})$$

$$\dot{m}_F(h_{1i} - h_{1e}) = \dot{m}_W C_W(T_{2e} - T_{2i})$$

$$C_W = 4.18 \text{ kJ/kg} \cdot \text{K}$$

$$0.8 \text{ kg/s } (310 \text{ kJ/kg} - 85.4 \text{ kJ/kg}) = 1.4 \text{ kg/s } (4.18)(\Delta T)$$

$$\Delta T = 30.7°\text{C}$$

Reference: NCEES *FE Reference Handbook* > Thermodynamics > Steady-Flow Systems, Selected Liquids and Solids table
Answer: C

104. The Moody diagram indicates that when $\text{Re} < 10^3$, the following equation should be used:

$$f = \frac{64}{\text{Re}} = \frac{64}{984} = 0.065$$

Reference: NCEES *FE Reference Handbook* > Fluid Mechanics > Moody, Darcy, or Stanton Friction Factor Diagram
Answer: D

105. Since the crack is on the exterior of the part, use a Y value of 1.1 (rather than 1.0). Convert 5 mm to m to calculate K as shown below.

$$K = Y\sigma\sqrt{(\pi a)}$$

$$K = 1.1(200 \text{ MPa}) \times \sqrt{(\pi(0.005 \text{ m}))}$$

$$K = 27.5 \text{ MPa} \times \text{m}^{1/2}$$

Reference: NCEES *FE Reference Handbook* > Materials Science/Structure of Matter > Fracture Toughness
Answer: C

106. Velocity ratio for the train: $m_v = \omega_{out} / \omega_{in}$

$0.5 = \omega_{out} / 24 \text{ rad/s}$

$\omega_{out} = 12 \text{ rad/s}$

Reference: NCEES *FE Reference Handbook* > Mechanical Engineering > Power Transmission

Answer: B

107. Specific humidity:

$\omega = \frac{0.622P_v}{P-P_v}$, where P_v is the pressure of vapor

$$0.0095 \text{ kg/kg} = \frac{0.622P_v}{101.3 - P_v}$$

$$0.96235 - 0.0095P_v = 0.622P_v$$

$$P_v = 1.523 \text{ kPa}$$

The dew point temperature (T_{dp}) is the saturation temperature of water vapor at a pressure of 1.523 kPa.

At $p_{sat} = 1.2276$ kPa, $T_1 = 10$°C

At $p_{sat} = 1.7051$ kPa, $T_2 = 15$°C

By interpolation:

$$\frac{T_{dp} - T_1}{T_2 - T_1} = \frac{P_v - 1.2276 \text{ kPa}}{1.7051 \text{ kPa} - 1.2276 \text{ kPa}}$$

$$\frac{T_{dp} - 10°\text{C}}{15°\text{C} - 10°\text{C}} = \frac{1.523 \text{ kPa} - 1.2276 \text{ kPa}}{1.7051 \text{ kPa} - 1.2276 \text{ kPa}}$$

$$T_{dp} = 13.09°\text{C}$$

Alternate Solution:
Refer to the ASHRAE Psychrometric Chart No. 1 (metric units) in the NCEES *FE Reference Handbook*. From the intersection of the dry bulb temperature and the humidity ratio, trace a horizontal line to find the dew point from the saturation curve.

Reference: NCEES *FE Reference Handbook* > Thermodynamics > Psychrometrics

Answer: C

108. Given that this is an orifice problem concerning flow, the following equation must be used:

$$Q = CA_0\sqrt{2g\left(\frac{P_1}{\gamma} + z_1 - \frac{P_2}{\gamma} - z_2\right)}$$

$P_1 - P_2 = 10\ \text{lb/ft}^2$, as the pressure difference is $10\ \text{lb/ft}^2$.

$z_1 - z_2 = 0$ ft, as there is no change in elevation. If the equation is rearranged, it makes more sense:

$$Q = CA_0\sqrt{2g\left(\frac{P_1 - P_2}{\gamma} + z_1 - z_2\right)}$$

The area (A_o), the cross-section of the entrance, is known. However, because the answer is in cubic feet per second (ft^3/s), $25\ \text{in}^2$ must be converted into square feet:

$$25\ \text{in}^2\left(\frac{1\text{ft}}{12\text{in}}\right)^2 = 0.17\ \text{ft}^2$$

You have A_o, $z_1 - z_2$, $P_1 - P_2$, $\gamma = 62.4\ \text{lb/ft}^3$, and $g = 32.2\ \text{ft/s}^2$. The orifice coefficient (coefficient of the meter) is all that is needed.

Use the Borda orifice table to obtain C (0.51). Plug the values into the equation:

$$Q = 0.51\ (0.17\ \text{ft}^2)\sqrt{2\ (32.2\ \text{ft/s}^2)\left(\frac{10\ \text{lb/ft}^2}{62.4\ \text{lb/ft}^3} + 0\ \text{ft}\right)}$$

$$Q = 0.28\ \text{ft}^3/\text{s}$$

Reference: NCEES *FE Reference Handbook* > Fluid Mechanics > Orifices

Answer: A

109. Point 1 is the compressor inlet, which is the intersection of P = 0.06 MPa and T = 40°C. Find the entropy and enthalpy.

$$s_1 = 2 \text{ kJ/kg} \cdot \text{K}, \qquad h_1 = 440 \text{ kJ/kg}$$

Point 2 is the compressor outlet. Since the compression process is isentropic, point 2 is the intersection of s = 2 kJ/kg·K and P = 0.4 MPa. Find the enthalpy.

$$h_2 = 490 \text{ kJ/kg}$$

Point 3 is the condenser outlet. Since the cooling process is isobaric, point 3 is the intersection of P = 0.4 MPa and the saturated liquid line. Find the enthalpy.

$$h_3 = 220 \text{ kJ/kg}$$

Point 4 is the evaporator inlet. Since the refrigerant remains at a constant enthalpy in the expansion valve, point 4 is the intersection of h = 220 kJ/kg and P = 0.06 MPa. Find the enthalpy.

$$h_4 = 220 \text{ kJ/kg}$$

$$\text{COP} = \frac{(h_1 - h_4)}{(h_2 - h_1)} = \frac{440 \text{ kJ/kg} - 220 \text{ kJ/kg}}{490 \text{ kJ/kg} - 440 \text{ kJ/kg}} = 4.4$$

Reference: NCEES *FE Reference Handbook* > Thermodynamics > P-*h* Diagram for Refrigerant HFC-134a, Common Thermodynamic Cycles table
Answer: B

110. In general: $P_1 \quad P_2 = -\gamma(z_2 - z_1) = -\gamma h = -\rho g h$

Note that the pressure of air is trivial compared to water.

$$P_A - P_B = \gamma_w(zB - zA)$$

$$P_B - P_C = \gamma_{\text{gas}}(zC - zB)$$

$$P_B = \gamma_{\text{gas}}(zC - zB) + P_C$$

Substitute:

$$P_A - [\gamma_{\text{gas}}(zC - zB) + P_C] = \gamma_w(zB - zA)$$

$$P_A - P_C = \gamma_w(zB - zA) + \gamma_{\text{gas}}(zC - zB)$$

Note:

$$P_{\text{gauge}} = P_A - P_C$$

$$P_{\text{gauge}} = \gamma_w(zB - zA) + \gamma_{\text{gas}}(zC - zB)$$

$$zB - zA = 15 \text{ cm} = 0.15 \text{ m}$$

$$zC - zB = 10 \text{ cm} = 0.10 \text{ m}$$

$$\gamma_w = \rho_w g = (1{,}000 \text{ kg/m}^3)(9.81 \text{ m/s}^2) = 9{,}810 \text{ N/m}^3$$

$$\gamma_{\text{gas}} = 0.6\gamma_w = (0.6)9{,}810 = 5{,}886 \text{ N/m}^3$$

$$P_{\text{gauge}} = (9{,}810 \text{ N/m}^3)(0.15 \text{ m}) + (5{,}886 \text{ N/m}^3)(0.10 \text{ m}) = 2{,}060 \text{ Pa}$$

Reference: NCEES *FE Reference Handbook* > Fluid Mechanics > Characteristics of a Static Liquid
Answer: C

You may have taken this practice test and studied the solutions,
but what else can you do to thoroughly prepare for exam day?

Keep practicing!

One of the best study tools we offer is our Practice Portal Pro, which contains a bank of practice problems and solutions that closely mimics the NCEES' computer-based test (CBT) experience. There, you will be able to see and work through problems in the same format and style as your exam. Detailed answers will also be provided so you can cross check your work and pinpoint where you may need improvement. Our goal with the Practice Portal Pro is to help study smarter, not harder, using retrieval-based learning strategies.

To gain immediate access to this comprehensive and time-saving study tool, download the free TotalAR app as explained at the front of this book. Use the app to scan the TAR code below and purchase the Practice Portal Pro (a $390 value).

This extremely useful tool will help you improve knowledge retention and gain a better understanding of the material with in-depth visual simulation.

www.schoolofpe.com